NUREG/CR–6312

Assessment of Fiber Optic Pressure Sensors

Manuscript Completed: March 1995
Date Published: April 1995

Prepared by
H. M. Hashemian, C. L. Black, J. P. Farmer

C. Antonescu, NRC Program Manager

Analysis and Measurement Services Corporation
AMS 9111 Cross Park Drive
Knoxville, TN 37923

Prepared for
Division of Systems Technology
Office of Nuclear Regulatory Research
U.S. Nuclear Regulatory Commission
Washington, DC 20555–0001
NRC Job Code W6315

ISBN-13: 978-1499576948; ISBN-10: 1499576943

This manuscript has been authored by a contractor of the U.S. Government under Grant No. W6315. Accordingly, the U.S. Government has a nonexclusive, royalty-free license to publish or reproduce the published form of this contribution, or allow others to do so, for U.S. Government purposes.

ABSTRACT

As a result of problems such as drift in nuclear plant pressure sensors and the recent oil loss syndrome in some models of Rosemount pressure transmitters, the nuclear industry has become interested in fiber optic pressure sensors. Fiber optic sensing technologies have been considered for the development of advanced instrumentation and control (I&C) systems for the next generation of reactors and in older plants which are retrofitted with new I&C systems.

This report presents the results of a six-month Phase I study to establish the state-of-the-art in fiber optic pressure sensing and describes the design and principle of operation of various fiber optic pressure sensors. This study involved a literature review, contact with experts in the field, an industrial survey, a site visit to a fiber optic sensor manufacturer, and laboratory testing of a fiber optic pressure sensor. The laboratory work involved both static and dynamic performance tests. In addition, current requirements for environmental and seismic qualification of sensors for nuclear power plants were reviewed to determine the extent of the qualification tests that fiber optic pressure sensors may have to meet before they can be used in nuclear power plants.

This project has concluded that fiber optic pressure sensors are still in the research and development stage and only a few manufacturers exist in the United States and abroad which supply suitable fiber optic pressure sensors for industrial applications. Presently, fiber optic pressure sensors are mostly used in special applications for which conventional sensors are not able to meet the requirements.

Fiber optic pressure sensors are typically more expensive than conventional pressure sensors and are not as readily available. Today, none of the traditional suppliers of nuclear-grade sensors supply fiber optic pressure sensors for safety-related applications in nuclear power plants. The susceptibility of fiber optic pressure sensors to nuclear radiation is a problem that may preclude the use of these sensors in high radiation environments of a plant.

Fiber optic pressure sensors provide EMI/RFI immunity, are generally more accurate than conventional pressure sensors, offer smaller size in both the sensor and the cabling, provide faster dynamic response, and allow multiplexing of signals which enables several process parameters to be measured at once. Fiber optic sensors can be used in flammable environments because they will not easily induce ignition. Also, the optical fibers themselves are chemically inert, which makes them suitable for corrosive environments and prevents them from affecting the process.

TABLE OF CONTENTS

TABLE OF CONTENTS
(Continued)

TABLE OF CONTENTS
(Continued)

LIST OF FIGURES

ACKNOWLEDGEMENTS

The cooperation of two fiber optic sensor manufacturers is gratefully acknowledged. Paroscientific, Incorporated of Redmond, Washington, contributed a fiber optic pressure sensor for the laboratory tests conducted in this project. Babcock & Wilcox (B&W), of Lynchburg, Virginia, arranged for a visit of Analysis and Measurement Services (AMS) personnel to B&W facilities in Alliance, Ohio, where fiber optic sensors are developed.

1. INTRODUCTION

Measurement of pressure in industrial processes is performed by a variety of sensors, most of which operate by converting the applied pressure to a mechanical movement. The mechanical movement is then measured by a displacement sensor and converted to an electrical signal. In conventional pressure sensors, the displacement is typically measured by electromechanical devices such as strain gages, differential transformers, capacitance sensors, and others. In fiber optic pressure sensors, the displacement is measured by altering light delivered by a fiber optic transmission system to the sensing element. The intensity or another characteristic of the return light is used to measure the displacement of the sensing element.

Fiber optic pressure sensors have several advantages over conventional pressure sensors. This includes high sensitivity, resistance to electromagnetic and radio frequency interference (EMI/RFI), reduced calibration requirements, fast response time, and electrical isolation. However, fiber optic pressure sensors have different failure mechanisms and failure modes than conventional pressure sensors and adequate data and experience do not currently exist on long term performance of these sensors in industrial processes. Therefore, substantial research is needed to establish the technical basis for the use of these sensors in nuclear power plants.

A number of industrial and government organizations including the Electric Power Research Institute (EPRI), National Aeronautics and Space Administration (NASA), National Institute of Standards and Technology (NIST), National Science Foundation (NSF), and the U.S. Department of Defense have interest in fiber optic sensing technologies. Research and development efforts are underway at these organizations as well as several universities to design and develop fiber optic temperature, pressure, vibration, and other sensors for industrial applications. Presently, fiber optic pressure sensors are used in limited applications in medical, aerospace, chemical, and automotive industries.

This report presents the results of a Phase I feasibility study to establish the state-of-the-art in fiber optic pressure sensing and determine the qualification requirements of these sensors for use in nuclear power plants.

Through a survey of manufacturers including a visit to a fiber optic sensor development facility, as well as the review of relevant literature and contacts with experts in the field, the operational characteristics, advantages, disadvantages, and failure modes of fiber optic pressure sensors were identified. These efforts were supplemented by an experimental evaluation of a fiber optic pressure sensor. A prototype pressure sensor was obtained from Paroscientific, Incorporated and laboratory tested to establish its static and dynamic capabilities at normal conditions. The results were compared to those of a conventional pressure transmitter to demonstrate the basic differences between the two sensors. In addition, the design and qualification criteria for the current generation of sensors in nuclear power plants were used to provide guidelines for fiber optic pressure sensors to be used in nuclear power plants.

2. BACKGROUND

Fiber optic sensors use light beams to transfer the sensed process parameter from the sensing element to the instrumentation where it is processed. Unlike the conventional sensors which typically vary the amplitude of an electrical signal in proportion to the changing process, fiber optic sensors can modulate the amplitude of the light, or use frequency, phase, or wavelength modulation. Each of these modulation techniques offers certain advantages and disadvantages which will be discussed in Chapter 6. In order to provide a basis for the discussions of the advantages and disadvantages of fiber optic sensors and describe the current state-of-the-art for this technology, some background information is presented below in this chapter.

2.1 History of Fiber Optic Sensors

Fiber optic technologies were pioneered by the telecommunications industry. Fiber optic communication lines offer many advantages over standard electrical cables such as increased bandwidth and lower attenuation. However, it was the aerospace industry that pioneered the use of fiber optics in process instrumentation. The major advantages of fiber optic instrumentation over conventional sensing systems for the aerospace industry include their small size, low mass, high accuracy and fast dynamic response capabilities. From here, fiber optic sensors expanded into other industries including the chemical, medical and automotive industries which required small and accurate sensors which could withstand the harsh environments to which they would be subjected. Although conventional sensors are still prevalent in most industrial applications, fiber optic instrumentation has found its way into many niche markets, mostly where conventional technologies have been inadequate. This typically involved harsh or electrically noisy environments or involved situations where size was a critical factor.

2.2 Principle of Operation of Fiber Optic Sensors

Figure 2.1 illustrates the basic components of a fiber optic sensing system. A transducer modulates a light signal according to the value of the process parameter being sensed. This modulated light signal travels through a fiber optic communication link to an interface unit. The communication link is typically in the form of a fiber optic cable or cables. Note that in some cases the transducer may be the cable itself. This intrinsic sensing is accomplished by allowing the process to alter the optical properties of the fiber core resulting in direct modulation of the light signal. The interface unit is used to either process the incoming light signal (ie, convert it to an electrical signal) or condition and multiplex it with light signals from other fiber optic instrumentation.

The basic operation of a fiber optic sensor involves a light source which provides light to a transducer. The transducer modulates the light that is then sent to an optical detector and then to the signal processing equipment. The light source for a fiber optic sensor is typically a light emitting diode (LED) or a laser. Both of these sources convert electrical

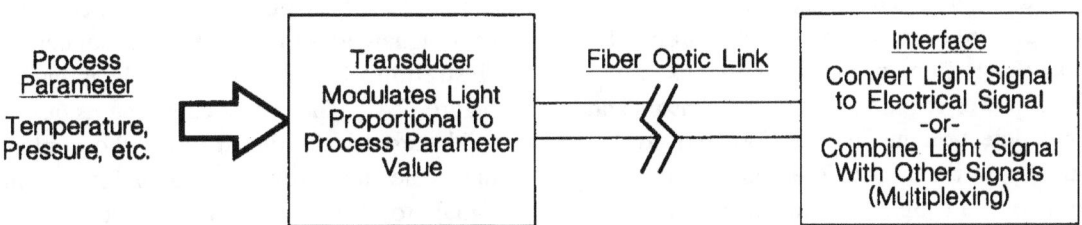

Figure 2.1 Illustration of the Basic Components of a Fiber Optic Sensing System

power into light with distinct spectral characteristics. For fiber optic systems involved in telecommunications, which involve long distance signal transmission, a laser is usually used as the light source because of its higher optical power output. However, fiber optic sensors typically utilize LEDs because of their availability, high reliability and low cost.[1] In a distributed control system containing many fiber optic process sensors, one light source can be shared among several sensors. This is accomplished using optical couplers as shown in Figure 2.2, which illustrates a common type of distributed system that uses time division multiplexing (TDM). The number of sensors which can be "daisy-chained" in this manner depends on the optical power requirements of the sensors and signal detection instrumentation and to the optical power output of the light source. An example of a simple coupling device is shown in Figure 2.3. In this figure, the collimating lens supplies parallel light beams to the beamsplitter which provides light to both of the receiving fiber cables through focusing lenses.

As seen in Figure 2.2, an optical detector receives the modulated light signals and converts them to electrical signals which are processed by electronic instrumentation. The optical detectors typically consist of a photodiode along with some conditioning circuitry which can amplify or buffer the electrical signal produced by the diode in order to interface properly with the signal processing devices.[1] The optical detector and the light source are selected by matching the characteristics of the light signals produced by the source and the characteristics of the light signals which can be effectively sensed by the optical detector. This allows for maximum efficiency in the fiber optic sensing system.

In some sensing systems, fiber optic technologies are used to transmit optical power to the transmitter. This optical power can be converted to electricity and used to power a conventional electrically-based transmitter. The modulated electrical signal is then converted back into a light signal for transmission to the processing electronics. This is commonly referred to as "Power by Light."

2.3   Fiber Optic Cables

The remaining component of a fiber optic sensor is the fiber optic cable which carries the light signals to and from the transducer. An illustration of a fiber optic cable is given in Figure 2.4. The main components are the solid core which runs through the center of the cable and provides a path for the light signal to travel and the cladding which surrounds the core. Although other components, such as inner and outer jackets, various coating materials and strength members, are typically included in fiber optic cables to provide added strength and life expectancy, they do not contribute to the transmission of the light signals and will not be discussed here.

2.3.1   Theory of Operation

As shown in Figure 2.4, the light enters the core and traverses the length of the cable by reflecting at the boundary between the core and the cladding. The total internal reflection of the light beam within the core is made possible by the difference in the optical properties between the core and the cladding. Note that the fiber shown in this figure is a multimode step index fiber. The propagation of light in other fiber types may differ and will be explained later.

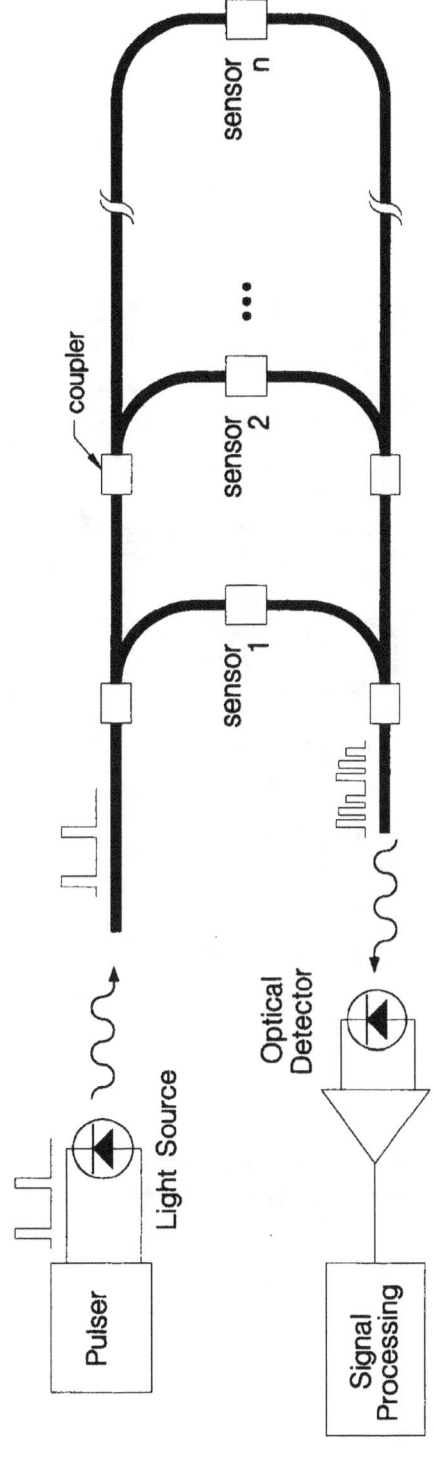

Figure 2.2   Illustration of Time Division Multiplexing in a Distributed Sensing System

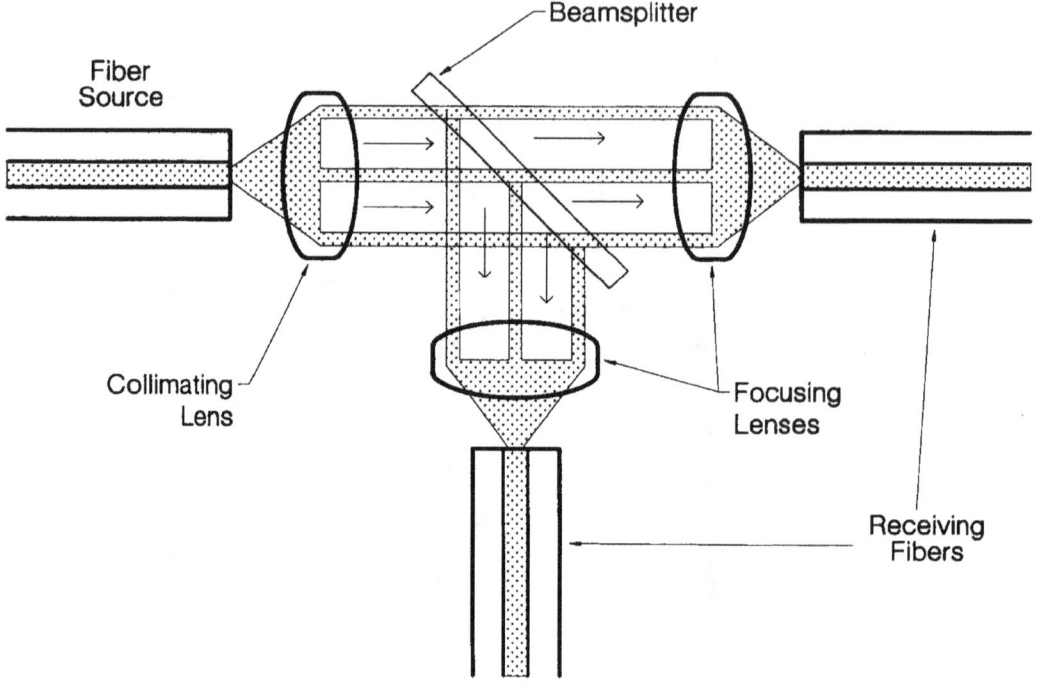

**Figure 2.3   Use of an Optical Beamsplitter to
Create Two Light Beams from One Source**

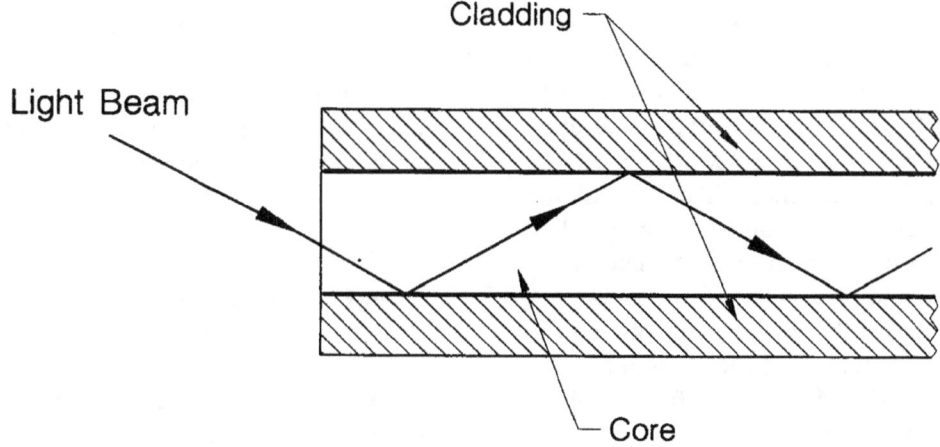

Figure 2.4   Propagation of a Light Beam Through a Fiber Optic Cable

Figure 2.5 shows three scenarios for a light beam striking the boundary between the core and cladding. The top of this figure shows the light beam striking the boundary at a small angle relative to the normal which is perpendicular to the boundary. This small angle (Θ) is called the angle of incidence and the light beam is referred to as the incident ray. According to Snell's law, if an incident ray encounters a border between two media with different optical properties, some or all of the light may enter the second medium, which is referred to as the refracted ray, while some or all of the light may be reflected back into the original medium. The amount of light which is refracted and reflected depends upon the velocities of light in the two media as well as the angle of incidence. The ratio of the speed of light in a vacuum to the speed of light in a particular medium is a dimensionless value known as the index of refraction (n). Snell's law gives the relationship between the angle of incidence and the angle of refraction (Φ) in terms of the indices of refraction of the two mediums as follows:

$$n_{core}\sin(\Theta) = n_{cladding}\sin(\Phi) \qquad (2.1)$$

where

$$
\begin{aligned}
n_{core} &= \text{index of refraction (core)} \\
n_{cladding} &= \text{index of refraction (cladding)} \\
\Theta &= \text{angle of incidence} \\
\Phi &= \text{angle of refraction}
\end{aligned}
$$

In the first case at the top of Figure 2.5, the small angle of incidence results in most of the incident ray entering the cladding at an angle greater than the angle of incidence. This is due to the fact that the index of refraction for the cladding is much smaller than the index of refraction for the core resulting in a greater angle of refraction. However, some of the incident ray is reflected back into the core. This is referred to as Fresnel Reflection and the angle of reflection is equal to the angle of incidence. Note that another possible scenario, in which the angle of incidence is 0 degrees, is not shown in this figure. In this case, all of the incident ray would enter the cladding along the normal with no reflection into the core. However, for all cases where the angle of incidence is greater than 0 degrees, some or all of the incident ray will be reflected back into the core.

The second scenario shown in Figure 2.5 represents an important situation. If equation 2.1 is solved for the sine of the angle of incidence, then:

$$\sin(\Theta) = [n_{cladding}/n_{core}]\sin(\Phi) \qquad (2.2)$$

If the angle of incidence is such that the ratio of the indexes of refraction for the core and the cladding is equal to the sine of the angle of incidence or:

$$\sin(\Theta) = n_{cladding}/n_{core} \qquad (2.3)$$

then:

$$\sin(\Phi) = 1 \text{ or } \Phi = 90° \qquad (2.4)$$

The angle of incidence (Θ) which causes this unique situation is called the critical angle (Θ_c). At this angle of incidence, as seen in the figure, all of the incident ray travels along the boundary between the core and the cladding after striking it.

The bottom of Figure 2.5 represents the ideal situation for a light beam traveling through a

Fresnel Reflection

AMS–DWG FOPO26A

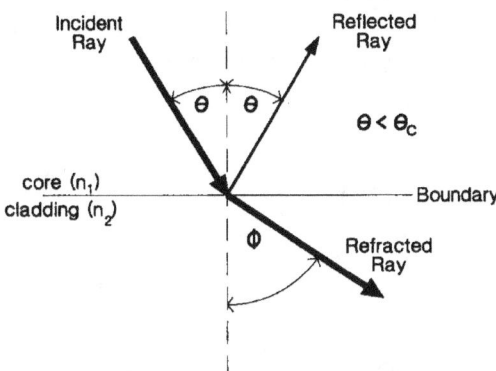

Critical Angle

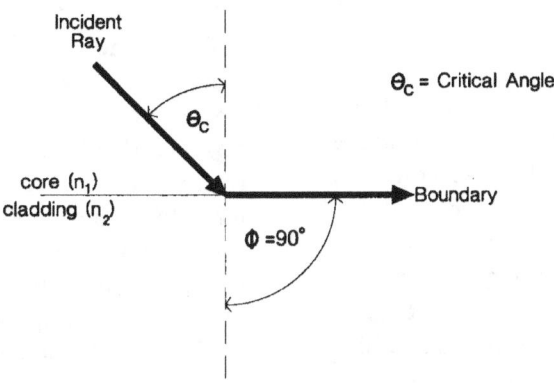

Total Internal Reflection

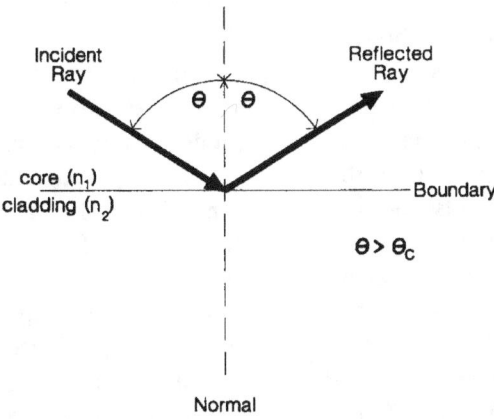

Figure 2.5   Three Scenarios for a Light Beam Striking the Boundary Between the Core and Cladding of a Fiber Optic Cable

fiber optic cable. In this scenario, the angle of incidence is greater than the critical angle or:

$$\sin(\Theta) > n_{cladding}/n_{core} \qquad (2.5)$$

therefore,

$$\sin(\Phi) > 1 \qquad (2.6)$$

which is not solvable for the angle of refraction (Φ). This means that no refraction occurs and all of the incident light beam is reflected back into the core at an angle equal to the angle of incidence. This situation, known as total internal reflection, repeats itself as the light beam travels the length of the cable.

Another important factor is the entry of the light beam into the cable. In order to promote total internal reflection within the core, the light beam must enter the cable at an appropriate angle, called the acceptance angle (Θ_a), so that it initially strikes the core to cladding interface at an angle greater than or equal to the critical angle. Light beams which enter the core at an angle larger than the maximum acceptance angle will be significantly diminished each time they strike the core to cladding interface resulting in reduced output power. Figure 2.6a illustrates this by showing the acceptance cone for light beams entering the core of the cable. Only light beams within this cone will propagate through the cable properly with the rest dissipating into the cladding. Note that the limits of the acceptance cone represent the angles at which the light source refracts into the core at an angle identical to the complement of the critical angle (90 degrees - Θ_c).[2] This refraction is due to the light beams entering the core from the air which has a different index of refraction. Numerical aperture (NA) is a term used to describe the size of the

acceptance cone for an optical fiber. It is equal to the sine of the maximum acceptance angle and is large for a fiber with a large acceptance cone (Figure 2.6b).

2.3.2 Cable Types

The materials from which fiber optic cables are manufactured vary greatly and depend on the requirements for the optical transmission being performed. Glass and silica are the most common materials for the core while sometimes plastic may be used. The cladding can also be made out of plastic or a grade of glass or silica different from that of the core. All-glass fibers offer the highest bandwidth and typically the lowest attenuation, or loss, although they are the most expensive type. These are typically used for long distance communications such as in the telecommunications industry. All plastic fibers have the highest attenuation but are the most rugged and easiest to handle. For fiber optic instrumentation, Plastic Coated Silica (PCS) fibers are typically used. PCS fibers have a glass core and a plastic cladding. They tend to have a higher numerical aperture than all-glass fibers and are less expensive.[1]

Depending on the characteristics of both the light source and the fiber optic cable itself, one or many modes of light may enter and propagate through the core. The number of modes distinguishes how many individual light beams are propagating through the core at the same time. The light beams which enter near the edge of the acceptance cone, and therefore strike the core to cladding interface more frequently, are referred to as the higher order modes. Those entering the core with a low acceptance angle are called the lower order modes. If only one mode can propagate through the core, then the fiber is called a single-mode fiber. If more than one mode is

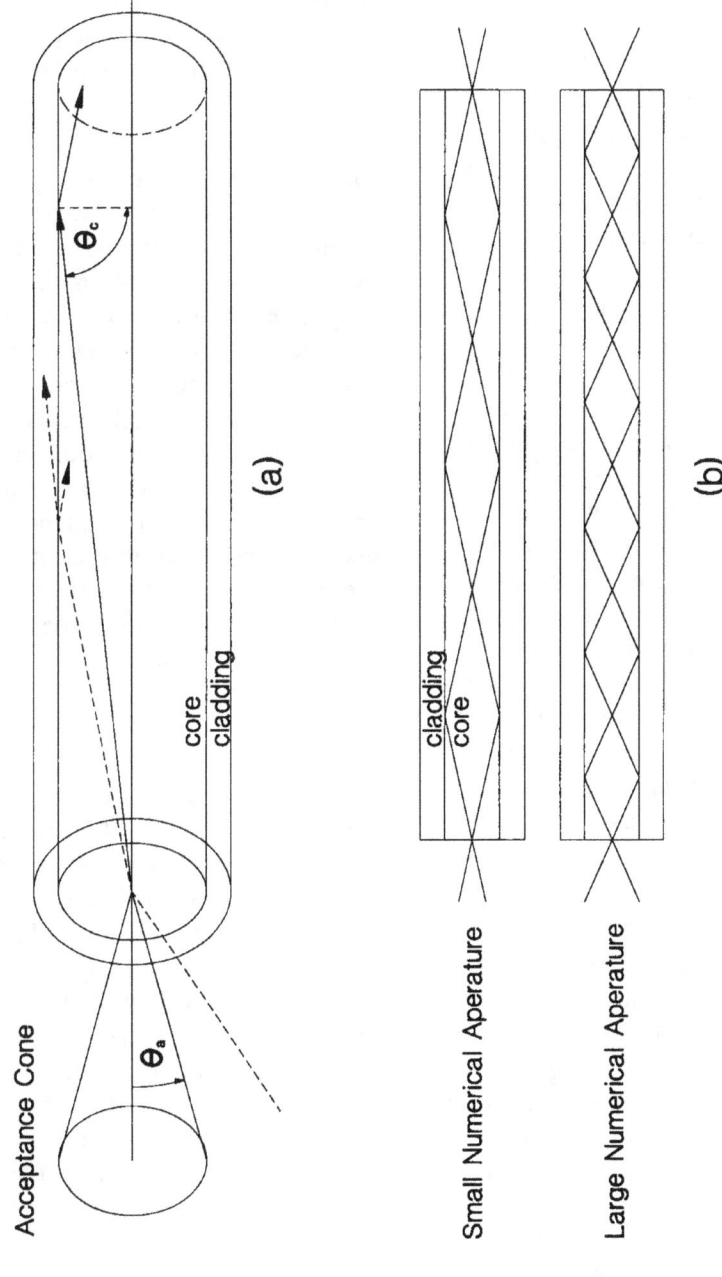

Figure 2.6 Illustration of the Acceptance Cone for a Light Beam Entering a Fiber Optic Cable and How Numerical Aperture Relates to the Size of the Acceptance Cone

allowed then it is known as a multimode fiber. Another important characteristic of the fiber is its refractive index profile. The two basic types of refractive index profiles are step index and graded index.

The combination of mode types and refractive index profiles gives many different types of fiber optic cables. Three of the fiber types are illustrated in Figure 2.7. The multimode step index fiber seen at the top of this figure has a distinct refractive index profile for both the core and the cladding making it a step index type. It is also seen in this figure that many different light modes can propagate through the core at the same time making it a multimode fiber. One of the major drawbacks of this type of fiber is known as modal dispersion. As seen in the figure, the light beams which have a large angle of incidence travel through the core over a shorter distance than those with a lower angle of incidence. Therefore, the different modes arrive at the end of the fiber at slightly different times. This limits the bandwidth of the fiber because adjacent light pulses must be separated by a time interval great enough to avoid overlapping, thereby reducing the speed at which information can be sent over the fiber.

In order to overcome the problem of modal dispersion, a second type of fiber, called a multimode graded index fiber, can be used. This fiber differs from the multimode step index fiber only in its refractive index profile. As seen in Figure 2.7, the cladding has a distinct refractive index but the refractive index of the core increases from the boundary between the core and the cladding to its maximum refractive index which is located at the center of the core. The result of the varying refractive index of the core is the propagation of light in an almost sinusoidal pattern as seen in the figure. Because of the increase in the velocity of the light beams with

decreased refractive index, the modes which are farthest from the center of the core travel faster than the other modes. This helps overcome the greater distance they have to travel and results in a lower modal dispersion than in the multimode step index fiber and therefore, a greater bandwidth.

The last type of fiber illustrated in Figure 2.7 is a single-mode step index fiber. This is identical to the multimode step index fiber except that only one mode of light can propagate through the core at one time. This is accomplished by decreasing the core diameter and the numerical aperture of the fiber or by using light beams with a longer wavelength. A combination of these factors can also be employed to attain single-mode operation. A single-mode fiber offers essentially no modal dispersion and therefore, the greatest bandwidth of the three fiber types. Although not shown in this figure, single-mode fibers can also employ a graded refractive index profile.

Single-mode fibers are the most efficient types of optical fibers in terms of their large bandwidth and low attenuation. However, they often require laser light sources and their small diameters, typically about 2 to 10 microns, require careful alignment when connecting them to the light source, couplers, etc.. Although this results in high costs to the end user, the advantages of this type of fiber make them desirable for telecommunications. Multimode graded index fibers, because of their higher bandwidth and lower distortion than multimode step index fibers, are often used for local area network (LAN) applications where the relative distances between the source and the detector are not as great as in telecommunications. For instrumentation purposes, multimode step index fibers are typically used because of their

Refractive
Index
Profile

Modal
Dispersion

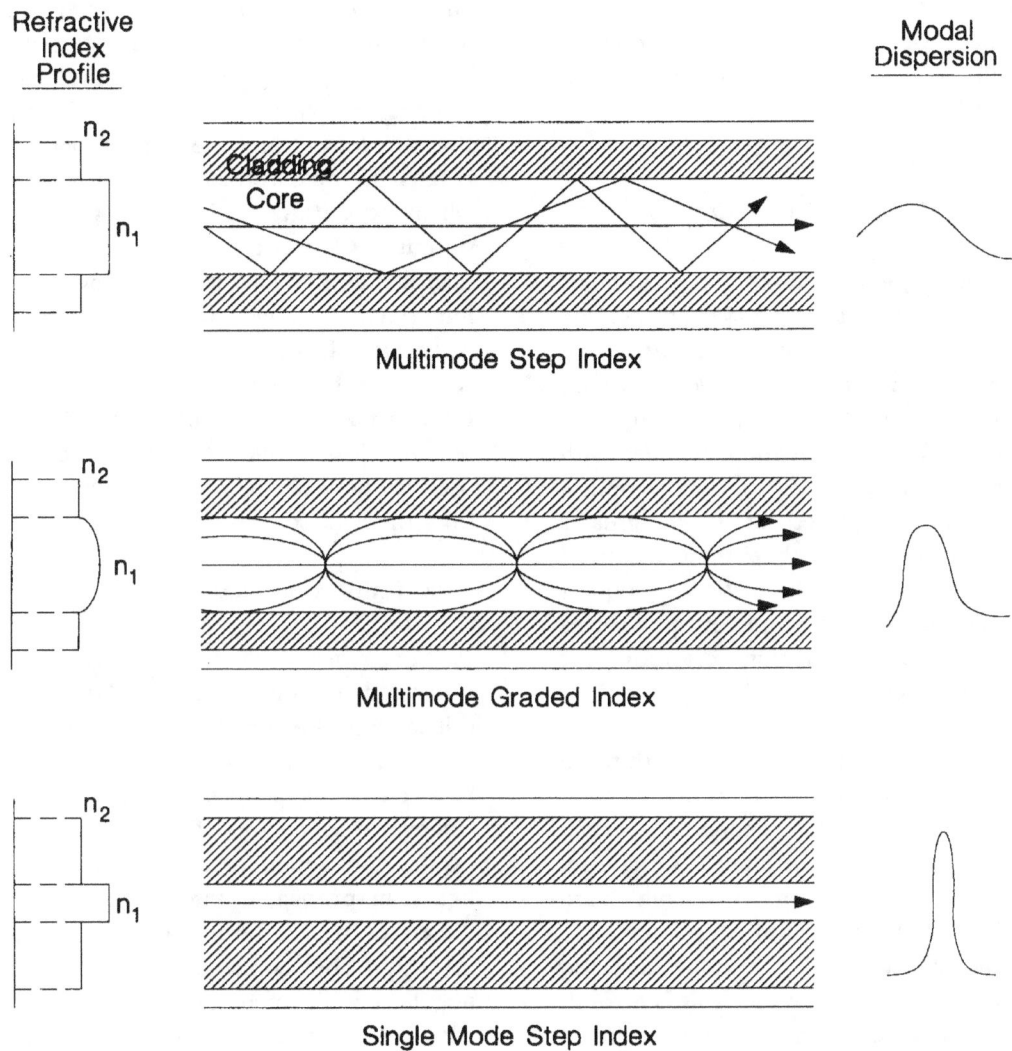

n_2

n_1

Cladding
Core

Multimode Step Index

n_2

n_1

Multimode Graded Index

n_2

n_1

Single Mode Step Index

Figure 2.7   Three Fiber Optic Cable Types and their Refractive Index Profiles as well as the Relative Amount of Modal Dispersion Associated with Each

low cost and easy connectivity. Their large core size, typically between 50 and 1,000 microns, makes them the easiest type of fiber to connect to and the modal dispersion is typically not a problem in industrial applications because of the lower required bandwidth and relatively small propagation distances.[1]

2.3.3 Optical Fiber Attenuation

One of the major advantages of optical fibers over electrical lines is their lower attenuation or loss. Also, optical fibers used for transmission of digital signals have a constant attenuation in relation to their frequency whereas electrical attenuation changes with frequency. These two factors allow optical repeaters to be placed at much greater distances from each other than electrical repeaters. However, attenuation in optical fibers still occurs. Losses in fiber optic cables can be due to absorption, scattering or excessive bending.[3]

Absorption in optical fibers is due to impurities in the core such as metal ions. These ions absorb light energy at certain wavelengths and turn it into heat. Of course, the lower the amount of impurities, which are a result of the manufacturing process, the lower the attenuation. Scattering can also cause losses in optical power. This is a result of inhomogeneities in the core which create localized changes in its refractive index resulting in scattering of the incident light, some of which may be reflected at an angle lower than the critical angle and lost in the cladding. Both absorption and scattering are caused by inherent properties of the fiber itself introduced during manufacturing.

The third source of attenuation is bending of the fiber. Figure 2.8 illustrates how macrobending results in attenuation of the optical signal in a multimode step index fiber. As seen in the figure, the normal, which is perpendicular to the core to cladding interface, changes throughout the bending region. Therefore, as some of the modes enter the bending region, they strike the core to cladding interface at angles of incidence (Θ) that are lower than the critical angle. This results in some refraction into the cladding which constitutes a loss in optical power output. The other type of bending is microbending. This is caused by ripples or imperfections in the core to cladding interface, or external forces exerted on the fiber. The reason for the attenuation is similar to that for macrobending. As will be seen in Chapter 5, bending losses due to external forces can be used to sense changes in the measurand for a fiber optic sensor.

2.3.4 Cable Connections

Proper connections between optical cables and other components, including other fibers, is a critical part of assembling a fiber optic system. The connection can be either permanent, referred to as a splice, or made by using removable connectors. Either way, the quality of the connection determines how well the light can propagate through it. Figure 2.9 illustrates three different causes of attenuation in optical fiber connections: end separation, angular misalignment, and lateral offset.[1] Of course, a particular connection may possess any combination of the three attenuation scenarios shown in this figure. The difficulty in obtaining a high quality connection is dependent on the tools and materials as well as the type of optical fiber used. The high numerical aperture of multimode step index fibers allows the most tolerance for interconnections while the single mode fibers, which have relatively small numerical apertures, require the highest precision.

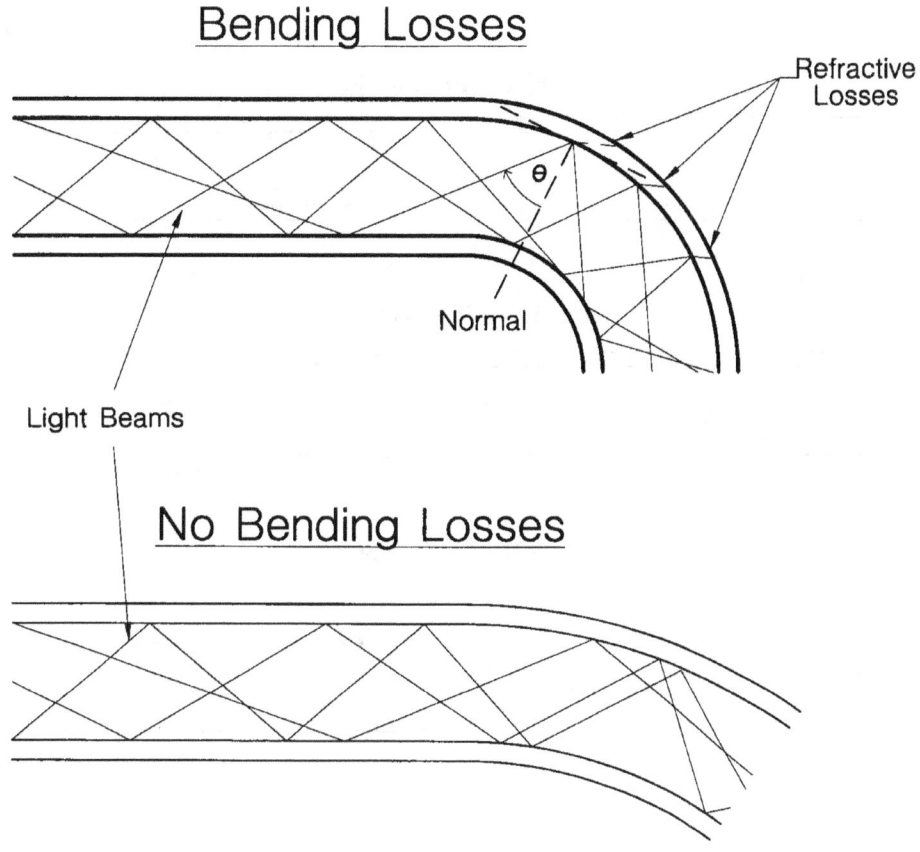

Figure 2.8   Attenuation or Loss of Signal Power due to Macrobending

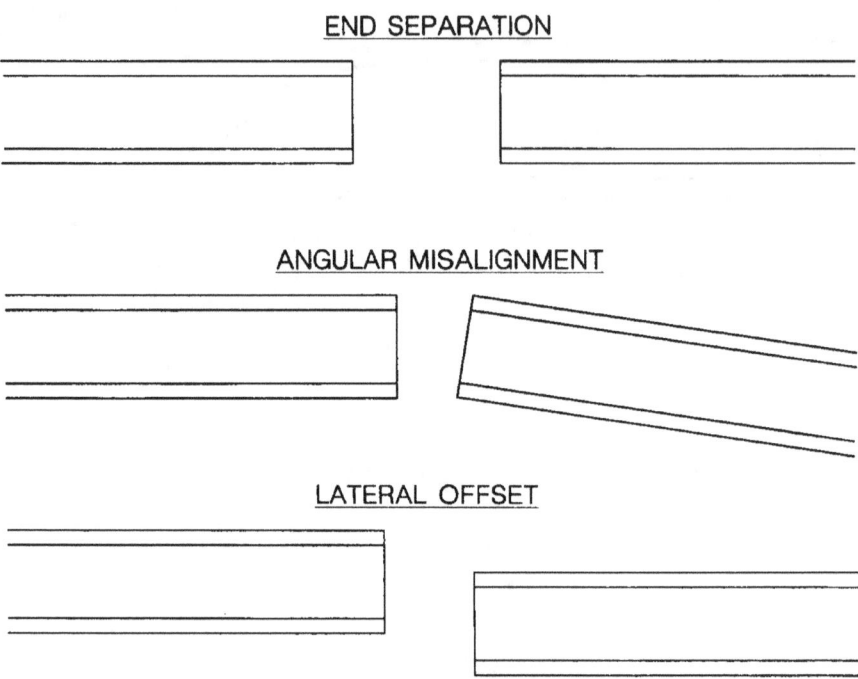

END SEPARATION

ANGULAR MISALIGNMENT

LATERAL OFFSET

**Figure 2.9   Illustration of Three Different Types of
Improper Mating Between Fiber Optic Cables**

3. RELATED RESEARCH

A number of organizations have been involved in research and development of fiber optic sensing technologies. A sample of these research and development efforts is summarized below.

3.1 NASA

The National Aeronautics and Space Administration (NASA) has sponsored and conducted several fiber optic sensing research projects. The topics of research have included electro-optical sensing and control in advanced aircraft and space systems,[4] fiber optics for aircraft communication, control, and sensing,[5] along with vibrational modal analysis.[6] In a NASA-Lewis funded Fiber Optic Control System Integration (FOCSI) project, fiber optic sensors were used to monitor pressure, temperature, flow, strain, and other parameters in the engine and airframe of an airplane. The fiber optic sensors were installed in parallel with existing instrumentation, and although the project funding has been completed, data is still being acquired and downloaded to a mainframe computer when the instrumented plane is flown. The fiber optic sensors installed in the airplane included a fiber optic pressure transducer developed by Babcock & Wilcox (B&W). This sensor is a dual-wavelength referenced microbend intensity-type pressure transducer.[7]

3.2 NIST

The National Institute of Standards and Technology (NIST) has solicited cooperative research efforts in applied technology in the area of optical electronics. The objectives of this initiative are the development of novel fiber optic sensing technologies, as well as methods to characterize the performance of optical and optoelectronic components.

The fiber optic sensors which have been solicited for development include pressure sensors as well as fiber optic sensors that measure voltage, current and other physical quantities. Since the nature of this research is applied technology, commercialization opportunities are emphasized.

3.3 EPRI-NSF

The Electric Power Research Institute (EPRI) and the National Science Foundation (NSF) have jointly allocated up to 3.5 million dollars in funds to support a research initiative of 9 to 12 awards in the area of Sensors and Sensor Systems for Power Systems and other Dispersed Civil Infrastructure Systems (SPSCIS).[8] In addition to electric power systems, the civil infrastructure systems as defined in the proposal solicitation include highways, railroads, waterways, and communications systems. However, most of the solicited topics address power generation and delivery.

The focus of the EPRI-NSF initiative is new approaches to sensor needs and issues, as opposed to traditional or conventional approaches. The primary goal of this initiative is to stimulate development of new sensing techniques, intelligent sensors, and optical sensors. Final proposal submissions have already been collected and research awards are planned for the fiscal year 1995.

3.4 EPRI

In addition to sponsoring research jointly with NSF, EPRI has initiated work at the Tennessee Valley Authority (TVA) and the Oak Ridge National Laboratory (ORNL) to investigate and evaluate fiber optic sensors for use in power plants. The purpose of these efforts, directed by EPRI's Office of Exploratory and Applied Research (EAR), is to advance the state of the art in optical sensing.[9] EPRI initiated an effort in 1992 at ORNL to identify and evaluate advanced pressure sensing concepts. Sixteen sensor technologies were considered by the researchers at ORNL. Fiber optic sensors were identified among the sixteen technologies as having the highest potential to both increase measurement accuracy and reduce maintenance cost.[10, 11]

Additional EPRI-funded work includes a wavelength-encoded pressure sensor developed at the United Technologies Research Center (UTRC) and a wavelength-encoded temperature and stress sensor developed at Mechanical Technology, Inc (MTI). MTI's sensor has been installed at the EPRI Instrumentation and Control Technology Center, located at TVA's Kingston Fossil Plant.[12]

Researchers at ORNL and Ohio State University have performed radiation testing on fiber optic cables at the University of Cincinnati's ^{60}Co irradiator (gamma irradiation) and the Ohio State University's Research Reactor (mixed gamma and neutron irradiation).[13] Radiation induced darkening of optical fibers and the resulting attenuation in fibers produced by various manufacturers was studied. The relationship between dose-rate, gamma/neutron radiation type, temperature, and optical power on the attenuation was also investigated as documented in Reference 18.

In order to help bring fiber optic sensors to the utility marketplace, and facilitate interaction between EPRI, power utilities, and optical sensor manufacturers, an Optical Sensing Manufacturers and Utilities Group (OPSM/UG) has been formed.[14]

3.5 U.S. Navy

The Naval Sea Systems Command (NAVSEA), the ship acquisition branch of the United States Navy, has developed a comprehensive effort to "take advantage of fiber optic sensors and put them on Navy ships."[15] NAVSEA is developing standards and specifications for the most promising fiber optic sensing technologies, including pressure sensing. Investigation into fiber optic sensor reliability and durability, as well as multiplexing is underway. Extensive reliability, durability, and failure mode analyses have been performed by the Carderock Division, Naval Surface Warfare Center (CDNSWC).[16]

4. CONVENTIONAL PRESSURE SENSORS

Pressure sensors currently used in the safety systems of nuclear power plants typically convert the mechanical displacement caused by the process pressure to an electrical signal with an amplitude proportional to the sensed pressure. This standard electrical signal is in the form of either a 4-20 or 10-50 milliampere (mA) current which is sent over two wires from the sensor in the field to the instrumentation in the control room area. This instrumentation also provides the power source for the sensor in the form of a direct current (DC) power supply.

The electronic instrumentation measures the incoming current signal and converts it to either digital information or to an analog signal which is used to drive indicators and the reactor trip logic and safety systems. Although there is much flexibility in the way the incoming current signal is processed, the use of an electrical signal to transfer the sensed pressure from the sensor to the electronic instrumentation is common to all conventional process sensors.

4.1 Pressure Sensing Elements

The pressure sensing element, which converts the measured pressure to a mechanical displacement, is typically one of three types: Diaphragm, Bellows, or Bourdon tube (Figure 4.1). These sensing elements are described below.

Diaphragm

A diaphragm sensing element consists of a circular plate fastened around the edges to a fixed surface. The change in pressure between the two sides of the diaphragm causes a mechanical displacement. Two diaphragms may be placed next to each other to form an elastic sensing element known as a capsule. The space between the diaphragms is typically filled with a fluid such as silicone oil.[17]

Bellows

A bellows-type sensing element uses many thin-walled tubes which form deep convolutions and is sealed at one end. The applied pressure moves a rod attached to the bellows which transmits the movement to the displacement sensor. This concept is similar to that of the diaphragm sensing element, but a more sizeable area is involved. A restraining spring is usually included with the bellows configuration to oppose the axial deflection. This spring can be adjusted in order to calibrate the transmitter.

Bourdon Tube

A Bourdon tube is a curved or twisted tube with an elliptical cross section which is sealed at one end. There are two common types of Bourdon tubes, the "C" type and the "spiral" type. The latter is made by winding the tube through several turns to amplify the displacement. The basis of operation of a Bourdon tube is that a closed end, coiled tube with a non-circular cross section will straighten out when pressure enters it. This results in an accurate and very repeatable sensing element.

4.2 Displacement Sensors

The conversion of the mechanical displacement caused by the sensing element to

Diaphragm

Bellows

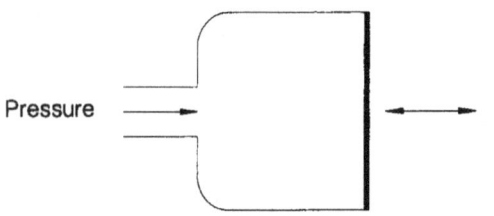

Pressure

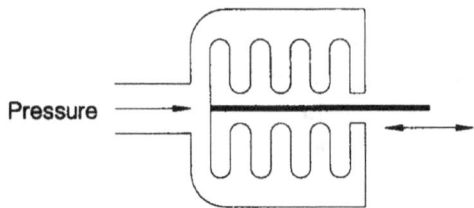

Pressure

Spiral-Type Bourdon Tube

C-Type Bourdon Tube

Scale

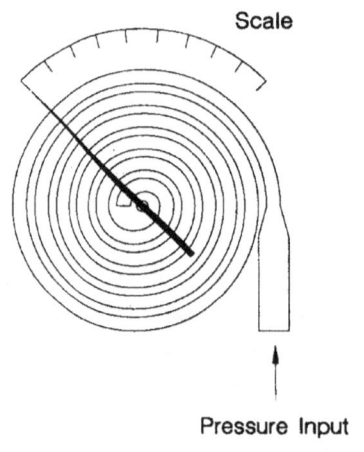

Pressure Input

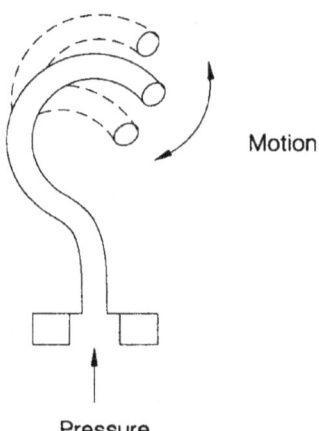

Motion

Pressure

**Figure 4.1   Types of Pressure Sensing Elements Used to
Convert Pressure to a Mechanical Displacement**

an electrical signal is necessary for the transmission of the pressure signal to the remote electronics. This is performed by displacement sensors. The most common displacement sensors are described below and are shown in Figure 4.2.

Strain Gauge

A strain gauge is manufactured such that a change in induced stress results in a change in resistance. As seen in Figure 4.2a, the wires in the strain gauge, which are typically very small, are commonly routed in a comb-like pattern to cover the maximum surface area and offer high measurement resolution. The changes in resistance are converted to electrical signals using a Wheatstone bridge circuit.

Capacitance Sensors

A capacitance sensor, illustrated in Figure 4.2b, consists of two fixed metal plates and a metal diaphragm in the middle which moves according to the applied pressure. The capacitances (C_1 and C_2) depend on the distances between the movable diaphragm and the fixed plates. This differential capacitance is used to modulate the electrical signal at the output of a bridge circuit similar to the one utilized by the strain gauges.

Inductance Sensors

Figure 4.2c shows one of several types of displacement sensors which utilize electrical inductance to measure the mechanical displacement. The movement of the armature changes the permeance, the magnetic equivalent of electrical conductance, of the magnetic flux path around the fixed magnet. This magnetic flux is induced by the exciter coil (A) on the left side of the magnet. The

voltage seen on the right side of the magnet is dependent on the distance of the armature from the magnet. The change in the size of the air gap results in a change in the electrical output induced in the sensing coil (B) on the right side of the magnet. Some inductance sensors use only one coil while others may change the relative positions of the two coils according to the amount of mechanical displacement.[17]

Differential Transformers

The basic operation of a differential transformer sensor is similar to that of an inductance sensor. A common type of differential transformer is the Linear Variable Differential Transformer (LVDT) shown in Figure 4.2d. When an alternating current (AC) electrical signal is applied to the primary coil it induces electrical signals in the two secondary coils (E_1 and E_2) with a phase difference dependent on the relative position of the transformer core which conveys the mechanical displacement. When the transformer core is in the center of the two secondary coils, E_1 and E_2 are out of phase by 180 degrees and, therefore, cancel each other out. As the core moves from the center, the relative phases change and a voltage (E_0), which is proportional to the mechanical displacement, is induced.

Potentiometer Sensors

Figure 4.2e shows a simple form of a potentiometer sensor in which a slide-wire resistor moves according to the sensed pressure. The difference between the output voltage and the input voltage is directly dependent on the position of the wiper. This is due to the change in resistance seen between the output terminals (A and C) as the wiper moves.

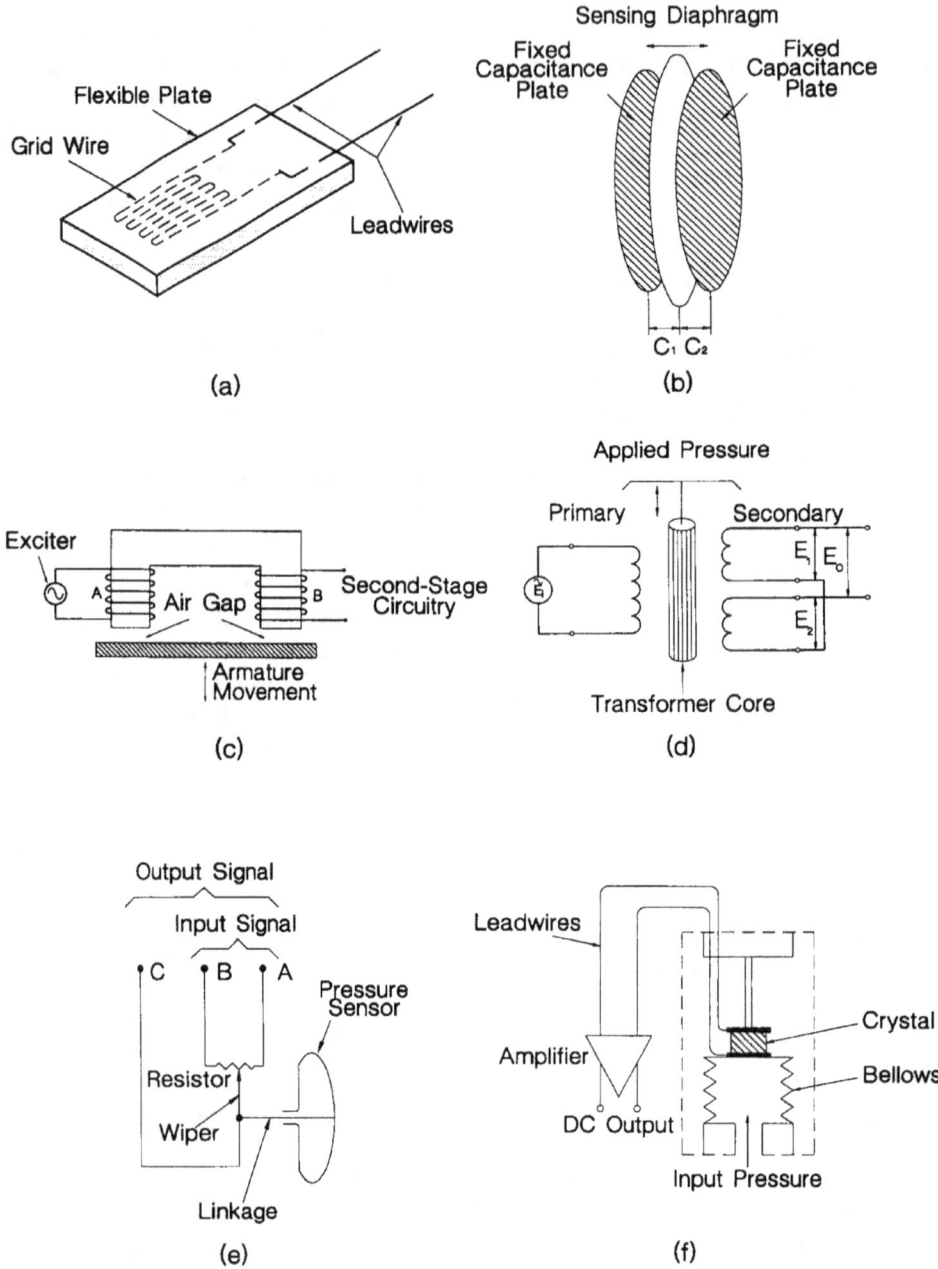

Figure 4.2   Six Types of Displacement Sensors which Convert Mechanical Displacement to an Electrical Signal

Piezoelectric Sensors

Piezoelectricity is the generation of an electrical signal due to strain, especially on a crystalline structure. Figure 4.2f illustrates the use of a quartz crystal as a piezoelectric sensor. The induced voltage from the crystal is very small and must be amplified as shown in this figure. This induced voltage is also direction-sensitive producing opposite voltage polarities for both tension and compression.

4.3   Smart Sensors

A new generation of process pressure sensors, known as smart sensors, is growing in popularity and finding its way into the nuclear industry. The major difference between smart sensors and other conventional pressure sensors is in how the transmitter modulates the electrical current signal according to the mechanical displacement. The generation of the analog signal produced by the sensing element and the displacement sensor can be accomplished through any of the means previously discussed. However, the smart sensor uses an Analog to Digital (A/D) converter to produce a digitized signal that represents the measured pressure (Figure 4.3).

The digital signal from the A/D can be conditioned according to the programmed coefficients which are stored in the transmitter's memory. These coefficients allow for zero and span adjustments, linearization, digital dampening, etc.[18] Information from internal sensors, which can monitor temperature and other environmental conditions which may affect the output of the transmitter, can be used to compensate for such effects in the digitized data. The adjusted data can be converted back to an analog current signal through a Digital to Analog (D/A) converter and sent through the electrical cables to the remote electronics. This allows the smart sensor to be used with existing instrumentation.

As seen in Figure 4.3, a communications unit can be placed across the current loop to enable control of the available features in the transmitter. The communications unit provides high input impedance to avoid affecting the current loop signal and uses high frequency AC signals to communicate with the transmitter. Because the AC signals do not interfere with the DC current signal, the communications unit is able to communicate with the transmitter while the sensing system continues to operate normally. Remote communications with transmitters in the field provide easy diagnostic testing, storage and recall of specific information about the transmitter, as well as error reporting which could not be done with conventional sensors.

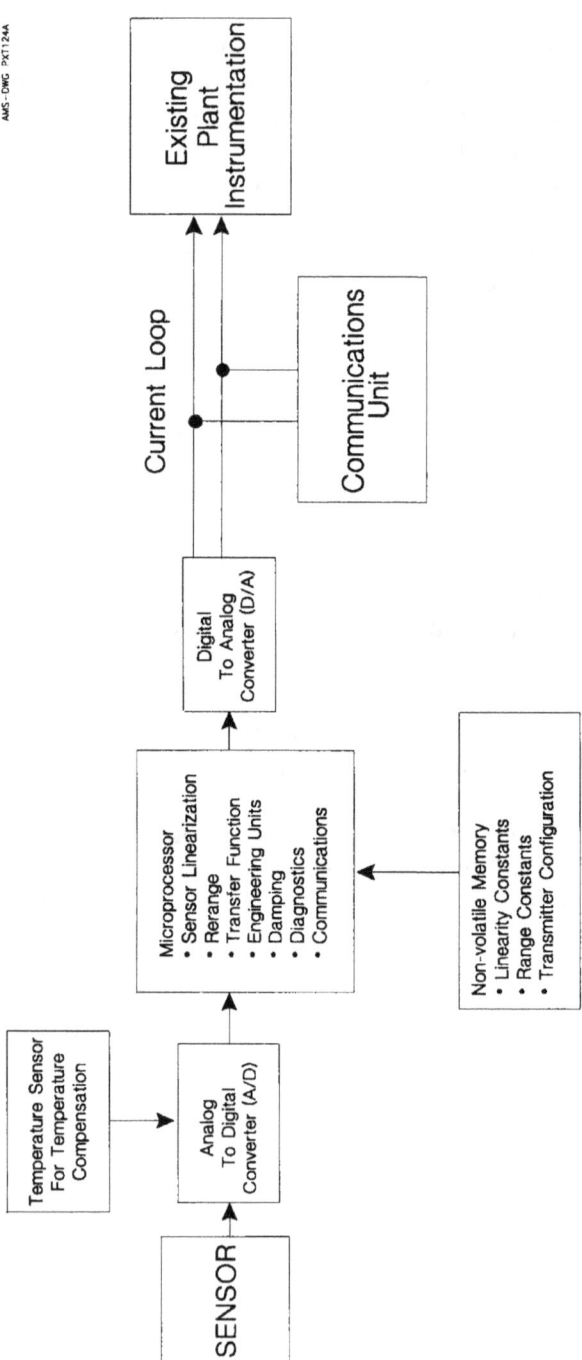

Figure 4.3   Block Diagram of the Components of a Typical Smart Sensor

5. FIBER OPTIC PRESSURE SENSORS

Two distinct methods may be utilized by a fiber optic sensing system to measure a process variable. These methods are referred to as extrinsic and intrinsic sensing. In an extrinsic sensor, the fiber optic cables are only used to supply light to and from an "off-fiber" transducer. The fiber optic cable may be viewed as strictly providing light to a "black box." After the black box has modulated the light signal with information about the measurand, a second fiber optic cable, or alternatively the original cable, transmits the information to a remote interface unit. In an intrinsic sensor, the modulation occurs inside the fiber. In this sensing mode, the measured property is allowed to deform the fiber which changes its optical properties resulting in modulation of the transmitted light.

In a fiber optic sensor, different characteristics or properties of the transmitted light can be modified to carry information about the measured variable. Fiber optic sensor designs may be divided into four main categories depending on the properties of the light signal that are modulated. These include intensity-modulated, phase-modulated, spectrum-modulated, and time and frequency-modulated sensors. In intensity-modulated sensors, which are also known as intensity-type sensors, the measurand affects the intensity, or brightness, of the light transmitted along a fiber optic cable. Phase-modulated, or interferometric, sensors encode the measurand in the phase difference between the light returning from a sensing optical path and the light from a reference optical path. Spectrum-modulated, or wavelength-encoding, sensors alter the spectral properties of the light. Other sensor types can modulate the frequency of the light signal or use an optical phenomenon known as fluorescence.

5.1 Intensity-Modulated Sensors

In intensity-type sensors, the light emitted from an optical source is carried along a fiber, its intensity is modified at the transducer and the light is returned to an optical detector. These sensors are analog in nature, as the light intensity detected is proportional to the measured variable. Intensity-modulated sensors can be classified as using one of three general modulation mechanisms: transmission, reflection, and microbending.

5.1.1 Transmissive Intensity Sensors

The transmissive concept is normally associated with intensity-modulated sensors in which the light is interrupted while passing through a break in the fiber optic cable. All of the transmissive sensors described below directly measure displacement or deflection of a diaphragm.

The simplest of the transmissive concept fiber optic pressure sensor designs is shown in Figure 5.1. A movable shutter connected to a flexible diaphragm is allowed to interrupt the light path between the light source and optical detector in proportion to the pressure applied to the diaphragm. Therefore, the intensity of the light received at the detector is related to the measured pressure.

A similar design, depicted in Figure 5.2, consists of both a fixed and a movable

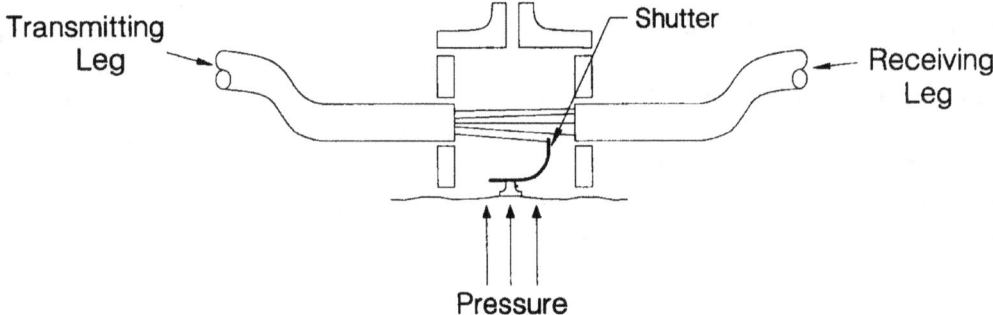

Figure 5.1   Illustration of a Movable Shutter Transmissive Intensity-Modulated Pressure Sensor

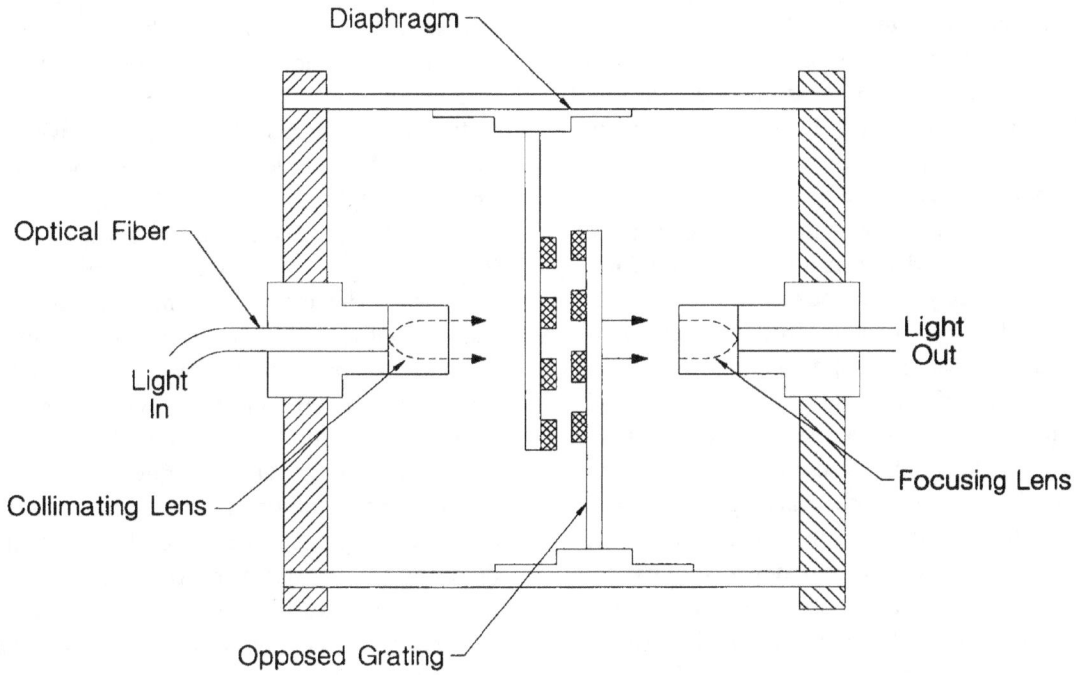

Figure 5.2   Drawing of an Opposed Grating Transmissive Intensity-Modulated Pressure Sensor

diffraction grating which modulate the light intensity reaching the detector. The fixed and movable gratings are located close to each other, so that with no displacement of the diaphragm, they are aligned and act as a single grating. As the diaphragm flexes under applied pressure, the movable grating blocks more of the light transmitted through the fixed grating. At a certain pressure, the movable grating is displaced by a distance equal to the grating spacing, and no light will be transmitted. As the grating spacing becomes smaller, the sensitivity of the sensor increases, achieving the highest sensitivity reported for a transmissive intensity sensor.[19] This sensor, as described, should only be used in an absolute pressure sensing configuration, as it is impossible to distinguish between positive and negative pressures across the diaphragm. Likewise, the applied pressure should not be allowed to reach the maximum pressure corresponding to zero light transmission. At this pressure, positive and negative pressure changes are indistinguishable. Use of a second set of gratings and probes would allow signal processing electronics to distinguish the direction of the change in deflection.

Another type of transmissive sensor involves displacement, either axially or radially, of a movable fiber optic segment to modulate intensity, as displayed in Figure 5.3. In these designs, the movable segment is mechanically connected to a diaphragm and the light beam travels from the fixed segment into the movable segment. The amount of light that reaches the movable segment is dependent on the amount of displacement. Radial displacement yields a more sensitive transducer and provides a relatively linear intensity reduction as the movable fiber goes from no displacement (full intensity) to displacement of one probe diameter (no transmission) from the fixed fiber. Axial displacement transducers have a greater range at the expense of less

sensitivity as the transmitted intensity is proportional to the inverse square of the displacement.

Frustrated total internal reflection (FTIR) is a modification of the transmissive concept (Figure 5.4). In this sensor, the ends of each fiber optic segment are polished parallel to one another at an angle to the fiber axis. An optically transparent liquid or gas with an index of refraction less than that of the optical fiber core separates the two fiber tips. One of the segments is fixed and the other is attached to a diaphragm. As pressure is applied to the diaphragm, the movable segment is displaced radially with respect to the fixed segment. Due to the interaction between the incident and reflected light in the movable segment, some light energy may be coupled into the fixed segment. This is due to a wave phenomenon known as evanescence. The amount of light coupled into the fixed fiber depends on its distance from the movable fiber. This sensor is one of the most sensitive of those employing transmissive intensity modulation.[19]

5.1.2 Reflective Intensity Sensors

The fiber optic cables in a reflective intensity sensor typically consist of several optical fibers in a bundle. Some of the fibers serve to transmit the light to the sensor, and the rest return the modulated light to the optical detector. The sensor consists of a reflective diaphragm or membrane that is allowed to deflect with the applied pressure. The light beam is "bounced" off of the reflective diaphragm and picked up by the receiving fibers (Figure 5.5). As the diaphragm is displaced, the intensity of the reflected light is modulated. This fiber optic sensor design enjoys several advantages, including non-contact measurement, simplicity and low cost.

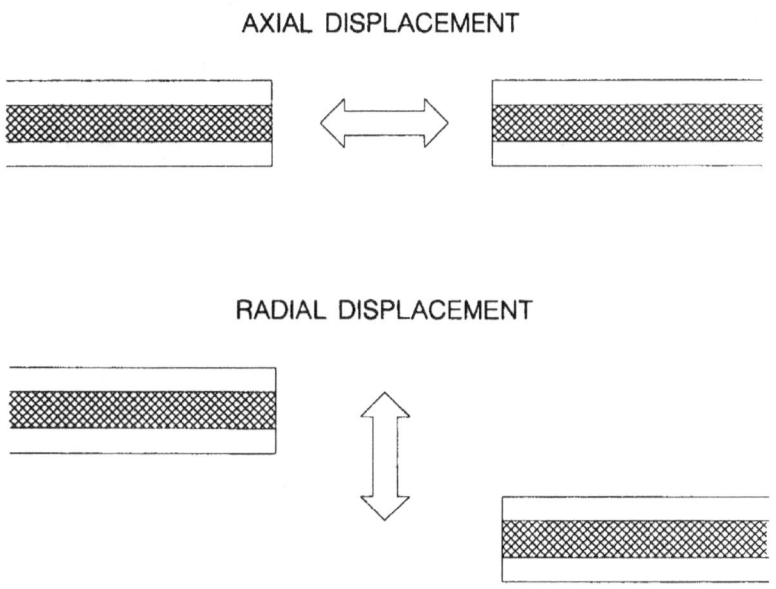

Figure 5.3   Illustration of Fiber Displacement as used in a Displacement Transmissive Intensity-Modulated Pressure Sensor

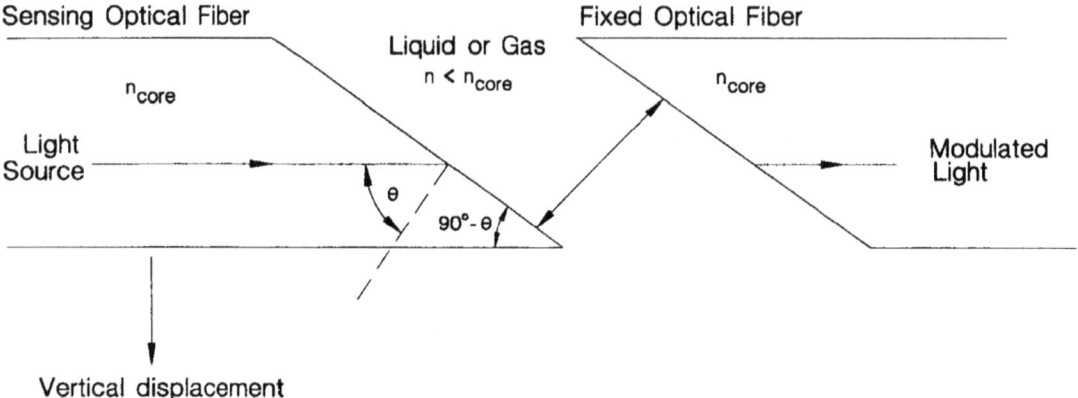

**Figure 5.4   Illustration of a Frustrated Total Internal
Reflection (FTIR) Pressure Sensor**

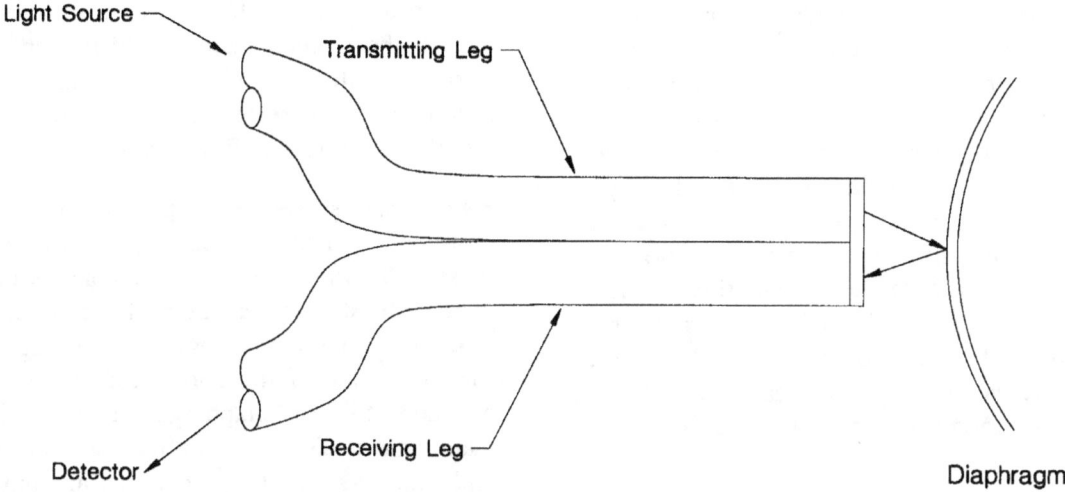

Figure 5.5   Example of a Reflective Intensity-Modulated Pressure Sensor

The geometric distribution of the transmitting and receiving fibers in the bundle greatly affects the sensitivity and linearity of the sensor response. Several arrangements are possible, as shown with their resulting response characteristics in Figure 5.6. The randomly distributed arrangement of fibers has been shown to provide a highly linear response, and is relatively insensitive to light source, fiber optic, and diaphragm orientations, as compared to other bundle arrangements.[20] The response characteristics of a particular bundle arrangement may be tailored to a specific application by use of a lens assembly attached to the end of the fiber optic bundle. This can be used to eliminate the front slope, or increasing portion, of each response curve giving the sensor a one-to-one relationship between distance and intensity. This eliminates the possibility of erroneous readings because of the two-to-one relationship of the original curve. The lens assembly can also be used to extend the range of the sensor.

Another reflective pressure sensor is the near total internal reflection (NTIR) sensor. This sensor, as shown in Figure 5.7, requires only one single-mode fiber, the end of which has been polished just slightly below the critical angle. The tip of the fiber is the sensor element, which is subjected to the process pressure. Light travels along the fiber, strikes the polished end, reflects to the mirrored surface, reflects back to the polished end, and is transmitted back along the fiber. Process pressure variations cause unequal changes in the refractive indices of the fiber and the surrounding medium. As the refractive indices change, the critical angle shifts. These shifts in the critical angle result in variations in the amount of light reflected back. This intensity-type sensor configuration possesses the advantage of being very small.

5.1.3 Microbend Sensors

Microbending, as described in Chapter 2, results in attenuation or loss due to some light beams refracting into the cladding. The higher order modes of light are the ones most affected by a microbend since they encounter the core to cladding interface at angles only slightly greater than the critical angle. Additionally, upon encountering a microbend, lower order light modes may be transformed into higher order modes which can be refracted into the cladding at the next microbend. A series of microbends can therefore lead to significant light losses.

Figure 5.8 shows a diaphragm pressure transducer containing a fiber optic microbend sensor. The microbend sensor consists of a multimode step index optical fiber which is squeezed between grooved or corrugated surfaces. One of the corrugated surfaces is attached to the diaphragm, and as the diaphragm is displaced, the fiber is squeezed and bent. As the fiber is bent, an amount of light proportional to the pressure applied to the diaphragm is lost due to microbending attenuation. In general, as the number of bending points on the corrugated surfaces is increased, and as the spacing between the corrugations is decreased, the sensitivity of the sensor is enhanced. The fiber optic cable in microbend pressure sensors is usually jacketed in a metallic or polymer buffer coating to protect the optical fiber from normal microbend stress, high temperature, and other environmental stressors. Also, this coating may extend the mechanical life of the sensor.

Figure 5.9 shows a different microbend-type sensor in which the fiber cable contains a continuous spiral-wound deformer element inside the protective jacket surrounding the cladding. The deformer element introduces microbending losses as lateral pressure is applied. The total attenuation can be related

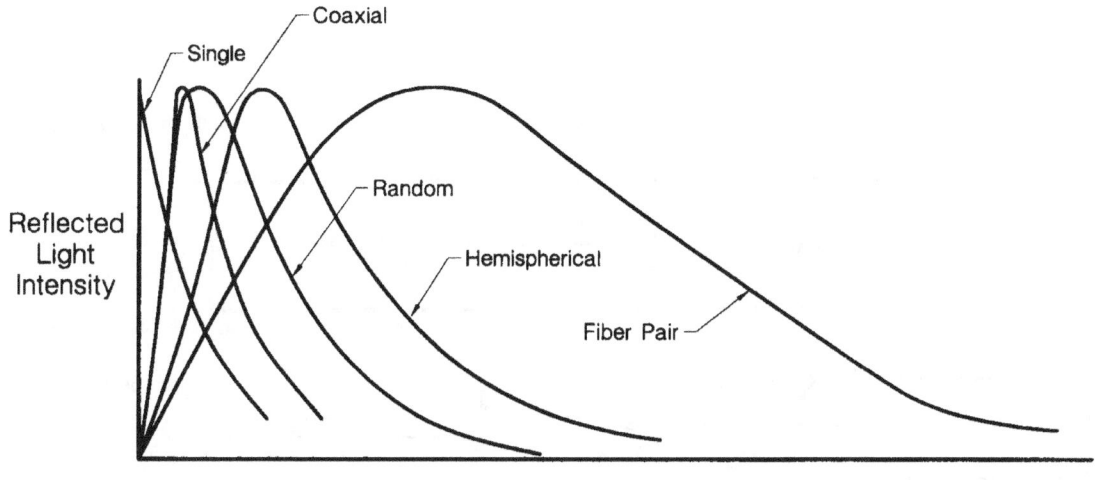

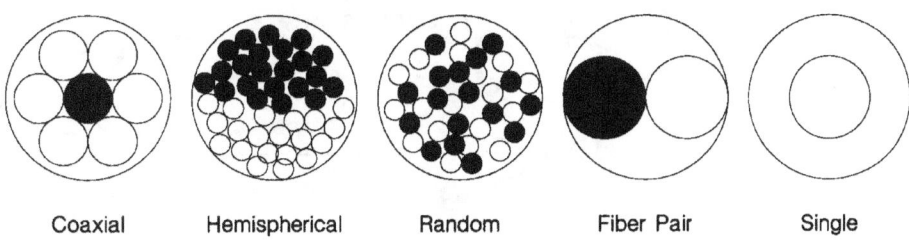

| Coaxial | Hemispherical | Random | Fiber Pair | Single |

**Figure 5.6   Five Types of Reflection Sensor Probe Bundle
Distributions and Associated Sensor Response Curves**

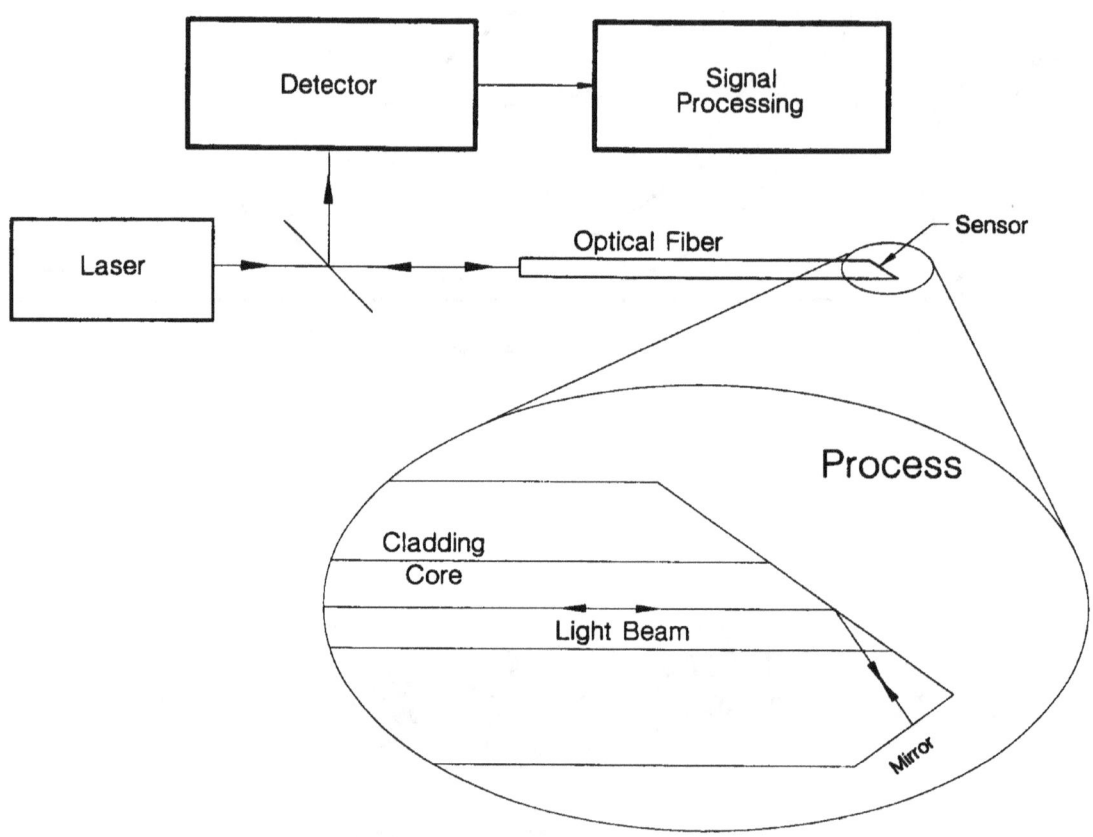

**Figure 5.7 Illustration of a Near Total Internal
Reflection (NTIR) Pressure Sensor**

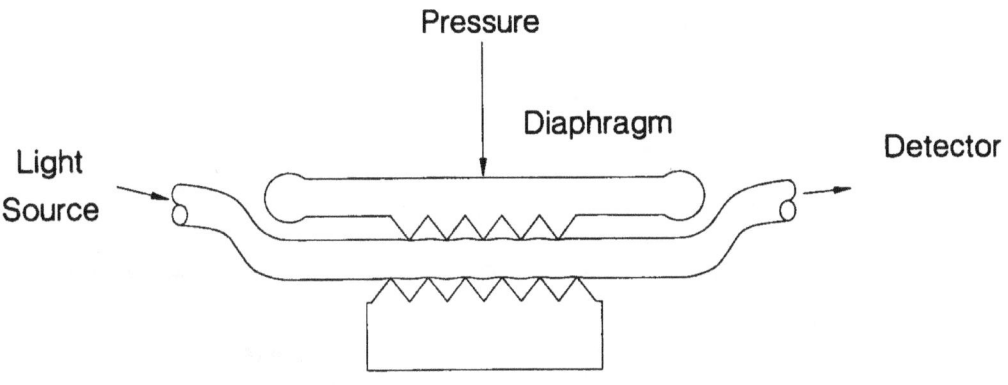

Pressure

Diaphragm

Light
Source

Detector

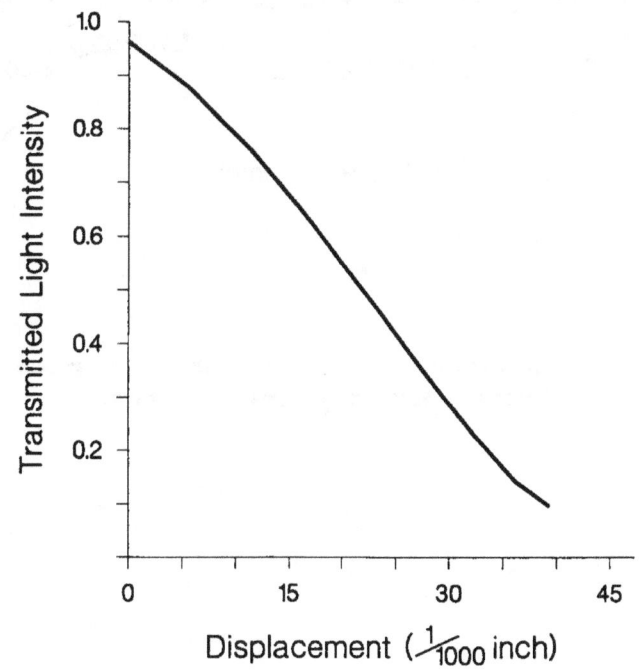

**Figure 5.8   Drawing of a Microbend-Type Intensity-Modulated
Pressure Sensor along with a Typical Response Curve**

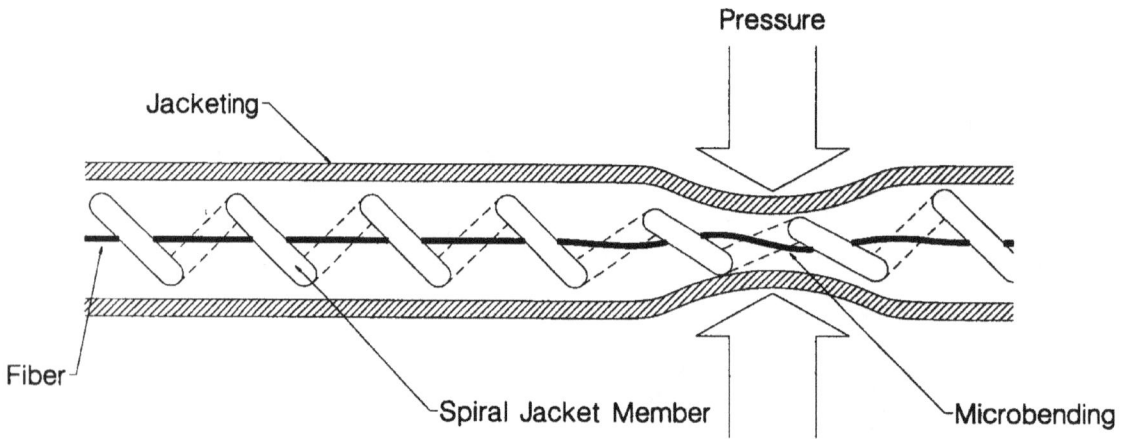

**Figure 5.9   Illustration of a Quasi-Distributed Microbend
Intensity-Modulated Pressure Sensor**

to the sum of the distributed pressures or alternatively, a technique known as optical time domain reflectometry (OTDR) may be used to determine the amount of pressure at specific locations.[21]

5.2 Phase-Modulated Sensors

Phase-modulated sensors use interferometric methods to sense the measured variable. Interferometry is the use of interference phenomena, based on the wave properties of light, to perform measurements. In phase-modulated sensors, changes in the measurand result in a phase difference between the modulated light and a reference light beam (Figure 5.10). Since the phase-modulated fiber optic sensors use interferometric measurement techniques, they are also referred to as interferometers.

There are four interferometric configurations: Mach-Zehnder, Michelson, Fabry-Perot, and Sagnac. The Mach-Zehnder, Michelson, and Fabry-Perot configurations may be utilized for configuration is chiefly used for gyroscopic applications and will not be covered here.

5.2.1 Mach-Zehnder Interferometer

The configuration of a Mach-Zehnder interferometric sensor is shown in Figure 5.11. The light source is split into a reference leg and a measurement leg. The measurement leg experiences both a length change and change in refractive index due to the pressure applied directly to the fiber. The two beams are then recombined and the phase modulation is detected by measuring the intensity of the recombined light.

The response and sensitivity of a Mach-Zehnder fiber optic pressure sensor are dependent on the type of buffer coating on the fiber optic cable. Metallic buffer coatings do not permit pressure-induced distortion of the fiber optic cable as easily as more flexible materials (e.g., plastics). This results in a reduction in sensitivity to applied pressure for metallic coated fiber optic cables, while plastic coatings result in increased sensitivity. In order to minimize undesired pressure sensitivity in the non-measurement regions of the interferometer (i.e. leads and reference fibers), metallic coated fibers may be used in those areas.

5.2.2 Fabry-Perot Interferometer

Fabry-Perot interferometric pressure sensors incorporate a resonance cavity, also referred to as an etalon, consisting of two partial reflectors on either side of an optically transparent medium. One of the reflectors, or mirrors, is attached to a diaphragm, and the cavity length is allowed to vary with the applied pressure. A schematic of the Fabry-Perot configuration is shown in Figure 5.12.

The light from the laser source impacts the first mirror, where some of the light is reflected back to the laser, while the rest of the light continues on to the second mirror. Some of this light is reflected back into the resonance cavity, and some is transmitted to the detector. The light which passes straight through the etalon without being reflected functions as the reference beam. Due to the high, but not perfect, reflectivity of the mirrors, some of the light which is trapped in the sensing cavity is bounced back and forth many times before it escapes to either the source or the detector. This results in a compounding of phase delays which increases the sensitivity of the Fabry-Perot sensor with respect to the other interferometric configurations.[19] However, experience has indicated that the majority of the light

Reference Fiber

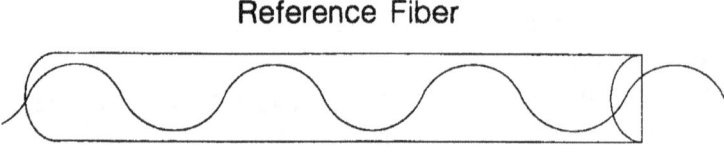

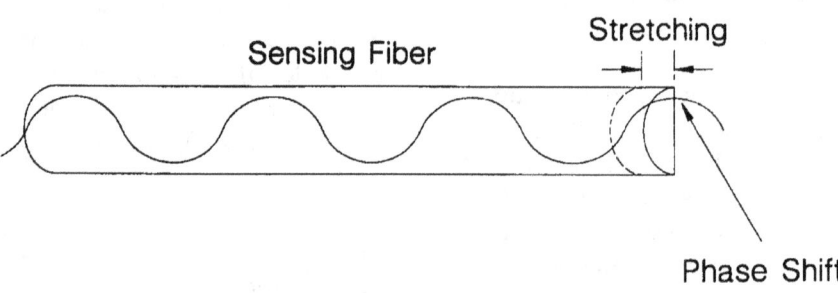

Sensing Fiber

Stretching

Phase Shift

Figure 5.10   Illustration of Phase Shift Due to a Change in the Length of the Fiber

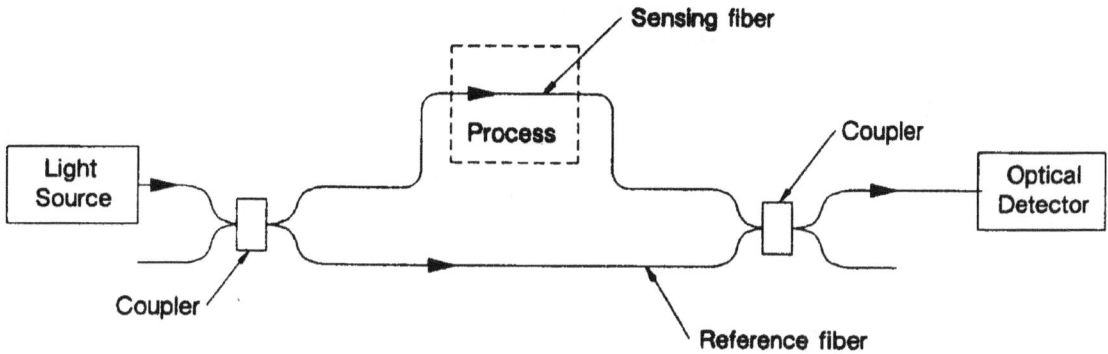

Figure 5.11   Drawing of a Mach-Zehnder Interferometric Pressure Sensor

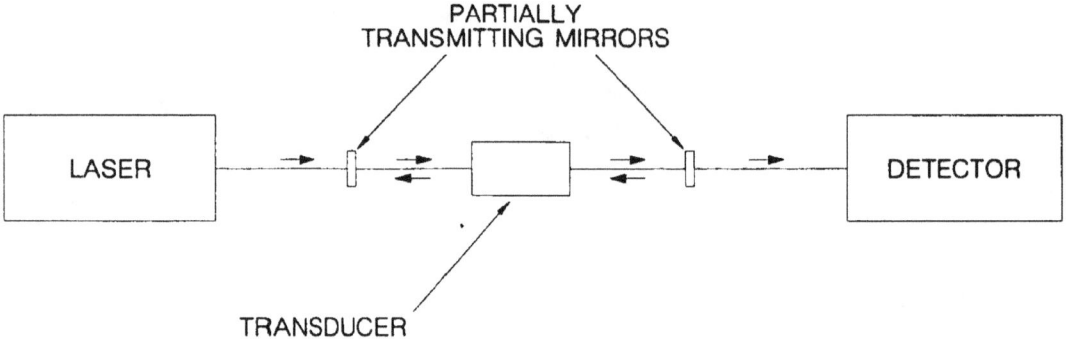

Figure 5.12   Drawing of a Fabry-Perot Interferometric Pressure Sensor

reaching the detector has either passed straight through the resonance cavity with no reflections, or has only experienced one extra cycle inside the cavity.[3,19,22,23] Another chief advantage of the Fabry-Perot configuration is that it only requires one fiber, and is therefore insensitive to intensity variations between the measurement and reference fibers.

The Fabry-Perot interferometer has at least two extrinsic configurations, as shown in Figures 5.13 and 5.14. Both designs include a coupler, as the light is required to reflect back in the direction of the laser source to reach the detector.

The extrinsic Fabry-Perot interferometer (EFPI) cavity depicted in Figure 5.13 consists of an air gap between the polished ends of the single-mode transmission fiber and a multimode fiber. The polished fiber ends serve as reflectors. The other end of the multimode fiber is shattered to minimize back-reflections of light which is not properly reflected at the polished end of the multimode fiber. As the process changes, the distance between the single-mode and multimode fibers changes. This results in a modulation of the phase difference between the reflected light beams.

The EFPI sensor shown in Figure 5.14 resembles a reflective intensity type sensor, in that the diaphragm serves as one of the mirrors which create the resonance cavity. The rest of the components in this sensor are configured exactly as in the previously described EFPI sensor and its operation is similar.

5.2.3 Michelson Interferometer

The Michelson interferometer configuration is illustrated in Figure 5.15. This interferometer is very similar to the Mach-Zehnder configuration, except that the sensing and reference legs are terminated with a reflective mirror. This results in the elimination of one coupler, but also introduces a significant disadvantage. In the Michelson interferometric configuration, the coupler feeds light back both into the detector and the laser. Feedback into the laser creates a source of optical noise which reduces the sensitivity of the interferometer.

5.3   Spectrum-Modulated Sensors

Spectrum-modulated or wavelength encoding fiber optic sensors rely on spectral attenuation or shifting to encode the measured process parameter. The detector must be able to determine optical intensities at different wavelengths to determine the value of the measurand.

Wavelength-modulated sensors incorporate a light source with specific spectral features. The light from the source is transmitted to a transducer, where wavelength-specific features are introduced (Figure 5.16). These features are introduced at one or more wavelengths related to the pressure applied to the transducer. Either a spectrometer, or alternatively, multiple filters and detectors are used to detect the spectral location of these features. In this figure, the process pressure alters the light intensity at a specific wavelength related to the process value.

5.3.1   Fabry-Perot Sensors

A typical wavelength encoding Fabry-Perot etalon sensor is shown in Figure 5.17. This sensor consists of a single fiber with a shallow cavity resonator at the tip. The light source to the cavity is provided by an LED with a narrow wavelength band. The interference of the reflected light beams results in modulation of the light intensity at specific wavelengths.

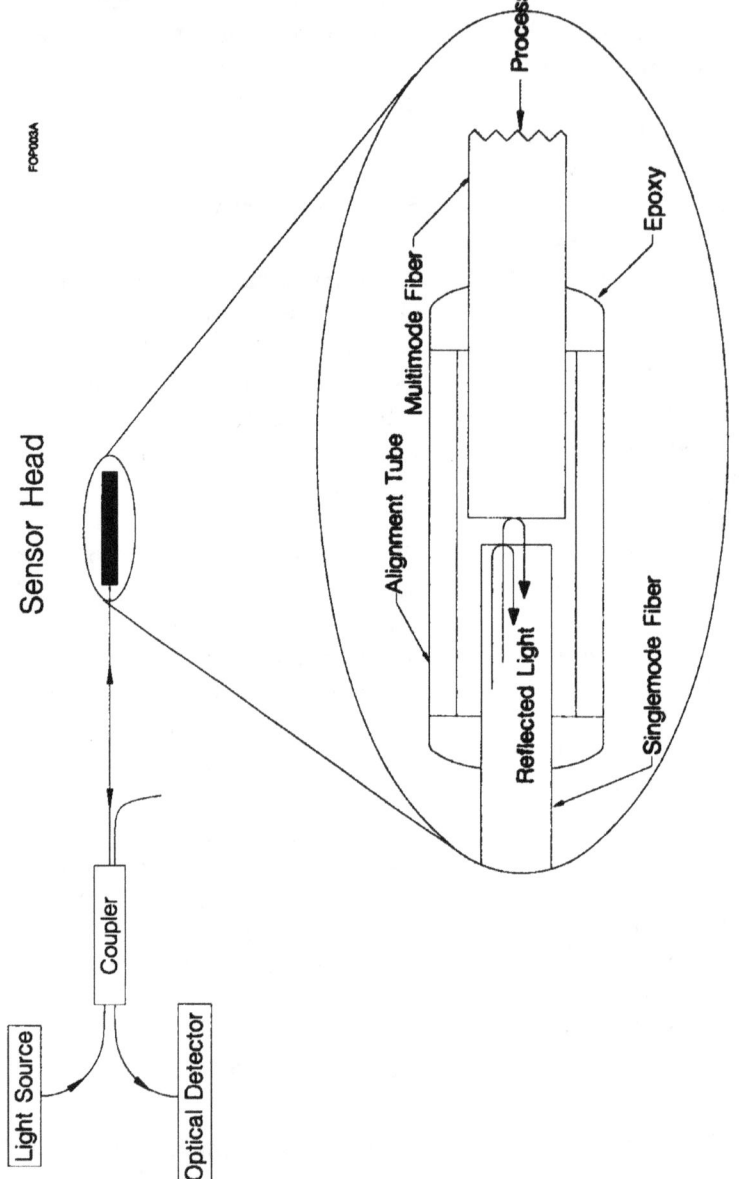

Sensor Head

FOP003A

Light Source

Coupler

Optical Detector

Process

Multimode Fiber

Epoxy

Alignment Tube

Reflected Light

Singlemode Fiber

**Figure 5.13   Drawing of an Extrinsic Fabry-Perot Interferometric (EFPI)
Pressure Sensor with a Multimode Reflector**

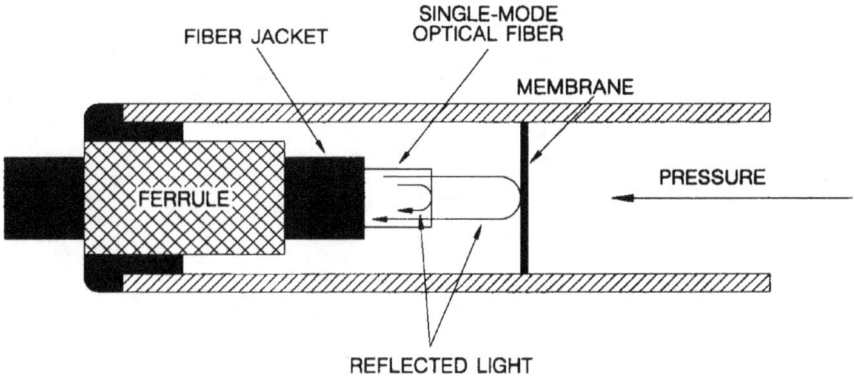

Figure 5.14   Drawing of an Extrinsic Fabry-Perot Interferometric (EFPI) Pressure Sensor with a Membrane Reflector

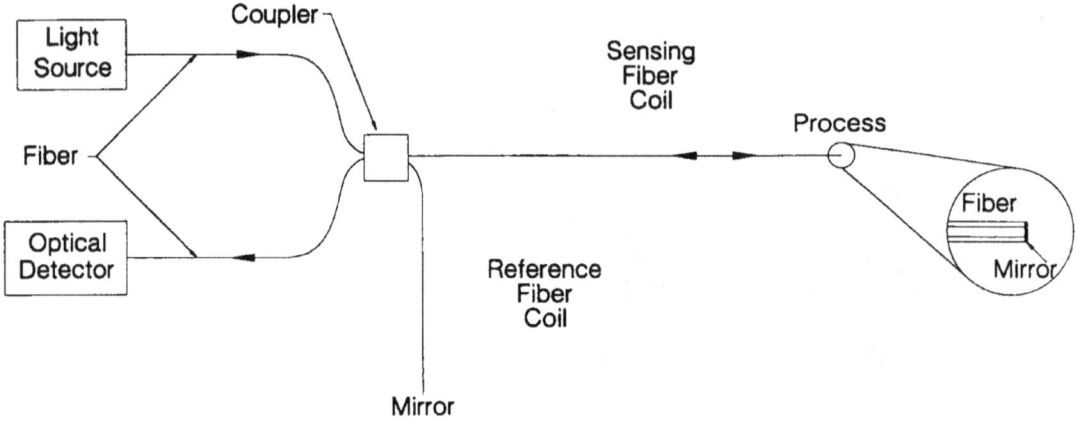

Figure 5.15   Illustration of a Michelson Interferometric Pressure Sensor

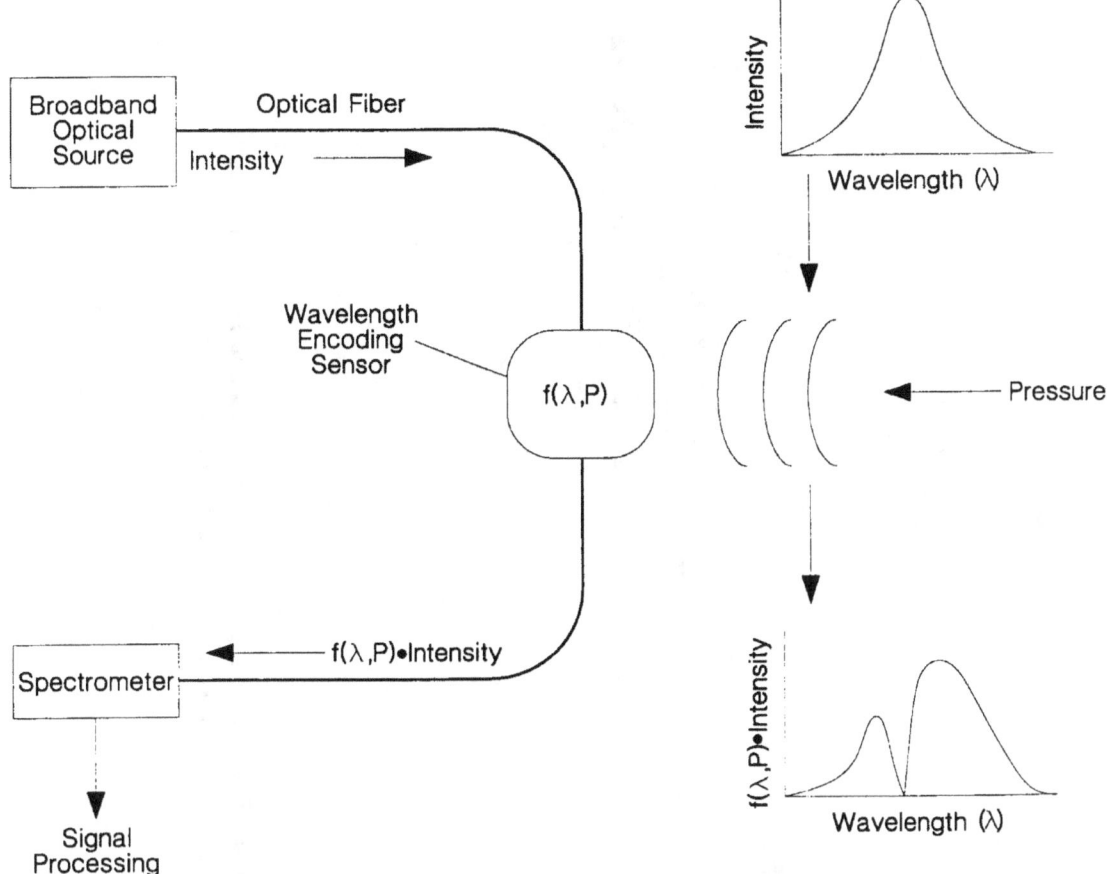

Figure 5.16   Diagram of the Wavelength-Encoding Pressure Sensing Concept

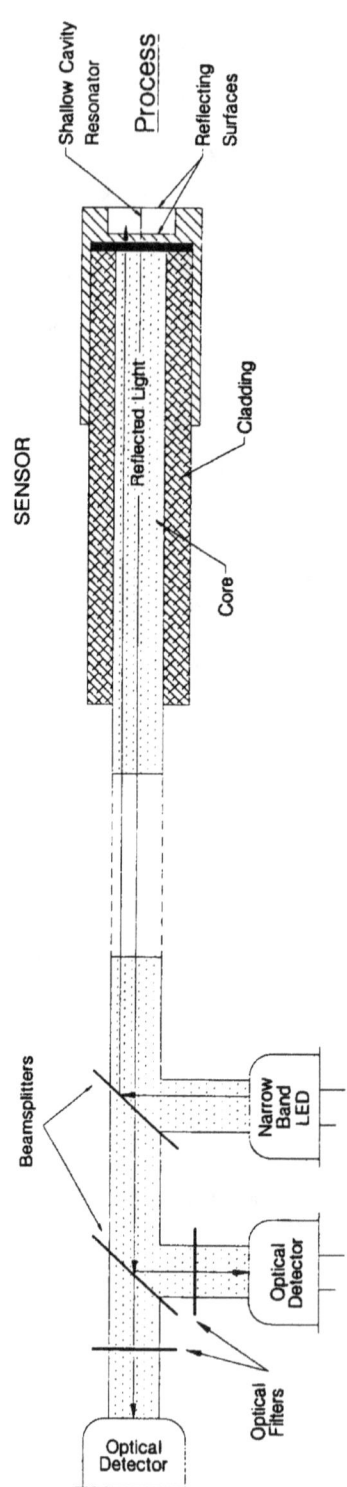

Figure 5.17   Drawing of a Fabry-Perot Etalon Dual-Wavelength Spectrum-Modulated Pressure Sensor

The modulated light is then split and subjected to two different optical filters which transmit distinct wavelength bands. The ratio of the intensities of the two different wavelength bands, or colors, is related to the external pressure. This sensor provides attenuation or loss compensation, assuming the loss mechanism is uniform ("grey") with respect to the two wavelength bands.

Another Fabry-Perot etalon based sensor utilizes a broadband LED along with a reference resonance cavity (Figure 5.18).[24] The reference cavity is used to detect the locations of a number of periodically spaced intensity minima in the output spectrum of the sensing resonance cavity. The locations of these minima are related to the applied pressure. Since light with a broad spectral band and multiple spectral features is being analyzed, this sensor is less sensitive than the previously described Fabry-Perot etalon sensor to wavelength-dependent attenuation and spectral shifts in the light source.

The light intensity detected at the optical detector is minimized when the separation of the reference etalon matches that of the sensing etalon. The separation of the reference cavity is varied by a piezoelectric translator (PZT). The PZT's voltage is controlled to minimize the electrical signal produced by the detector. The reference cavity's mirrors are metal coated and the impedance of the parallel-plate capacitor formed by the cavity is measured and related to the process variable.

5.3.2 Absorption Spectral Attenuation Sensors

The sensor shown in Figure 5.19 uses the wavelength-dependent filtering of an absorbing glass modulator to modulate the spectrum of the light signal. The modulator glass is placed between two segments of a transmission loop and is displaced radially with respect to the segments as a diaphragm is deflected with applied pressure. A broadband source is used to provide the light that is transmitted from one fiber segment to the other. Some of the light exiting the transmitting segment encounters the glass modulator and is filtered as it propagates to the receiving segment. The rest of the light is unmodulated as it enters the receiving segment. The received light is sent to two optical detectors after being filtered at two separate center frequencies. The ratio of the two detected intensities is related to the pressure applied to the diaphragm.

5.3.3 Diffractive Wavelength-Modulated Sensors

The diffractive property of reflective gratings may be used to provide wavelength modulation for fiber optic pressure measurement. Figure 5.20 shows how the grating spacing may be varied with mechanical displacement to provide narrowband light at a specific output angle. This narrowband light will then vary in wavelength as the grating is horizontally displaced. As shown in Figure 5.21, a spectrometer, or similar device, is used to detect the spectral properties of the returned light.

5.4 Time and Frequency Modulated Sensors

Some fiber optic pressure sensors time-code the measurand and are essentially independent of attenuation effects. Sensors which fall into this category either use a frequency signal generated by a quartz crystal resonator or a fluorescent decay rate to transmit the measured pressure.

5.4.1 Resonating Element Quartz Transducer

Sensors utilizing a resonating element quartz transducer operate by modulating the resonant

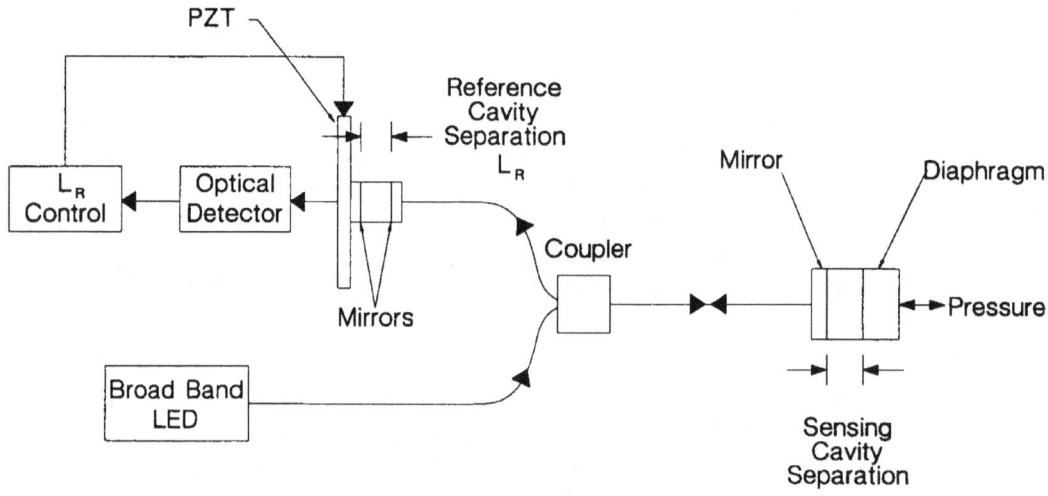

**Figure 5.18   Drawing of a Fabry-Perot Etalon Spectrum-Modulated
Pressure Sensor Using a Reference Etalon**

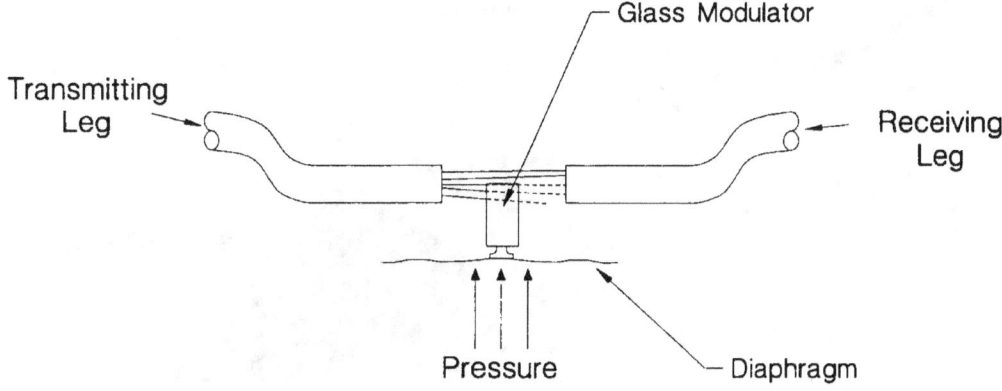

Figure 5.19   **Illustration of an Absorbing Glass Spectrum-Modulated Pressure Sensor**

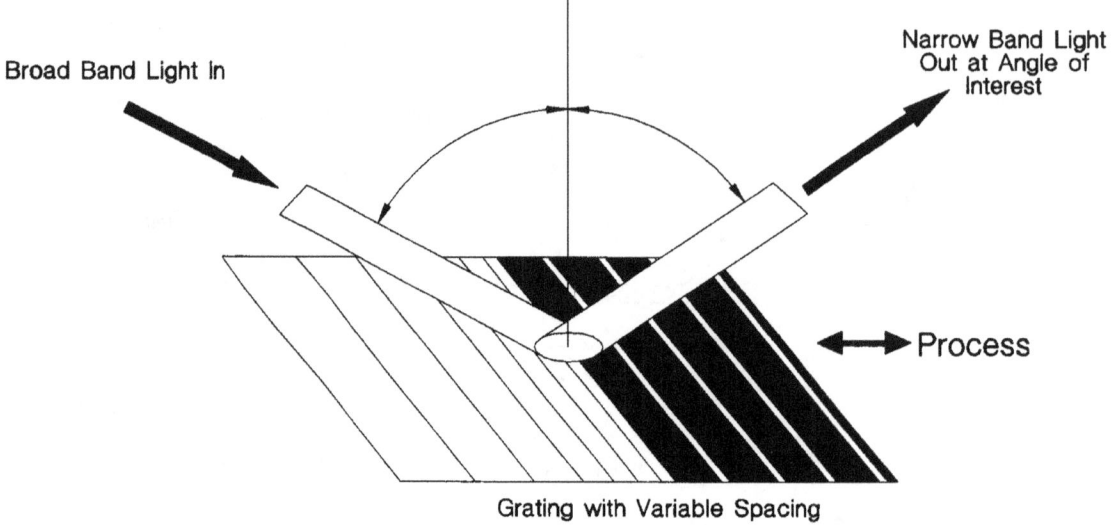

Figure 5.20   Illustration of the Diffractive Property of a Reflective Grating

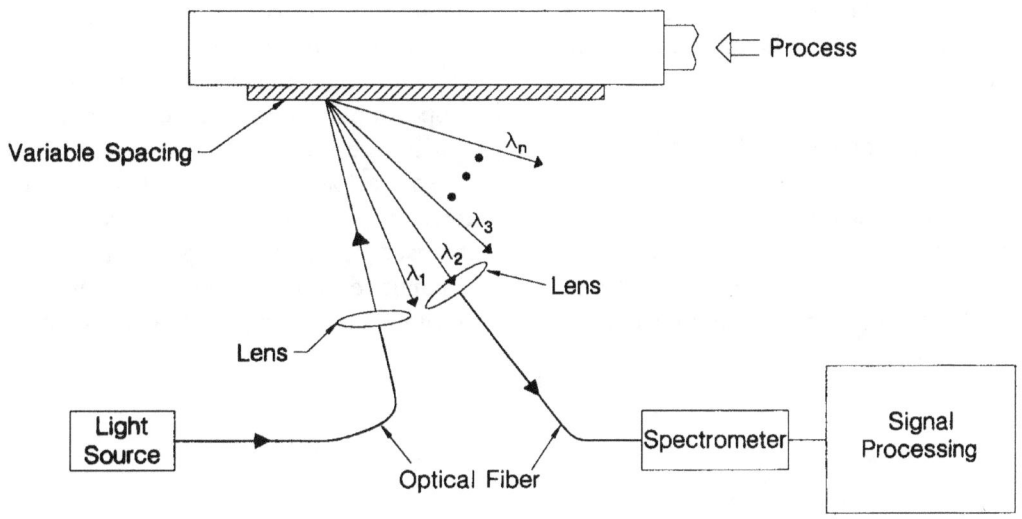

Figure 5.21   Drawing of a Reflective Configuration of a Diffraction Grating Spectrum-Modulated Pressure Sensor

frequency of load-sensitive quartz crystals. This change in resonant frequency can be used to modulate the light beam. The pressure sensing element is designed to minimize environmental effects, and a similar quartz crystal temperature element can be used to provide additional thermal compensation. The pressure and temperature frequency information is optically transmitted back through a separate fiber optic cable to the optical detector and signal processing electronics. A fiber optic pressure sensor utilizing this technique will be described in Chapter 10.

5.4.2   Fluorescent Decay Sensor

A fiber optic pressure sensor which uses fluorescent decay-rate modulation is shown in Figure 5.22.[25]   This sensor design incorporates a single multimode optical fiber to supply light to the sensor and to transmit the modulated signal back to the optical detector. The end of the fiber is attached to a diaphragm causing the fiber tip to move vertically from one fluorescent disc to another as the diaphragm is deflected with the applied pressure. The sensor contains two different neodymium-doped glass fluorescent discs characterized by distinct fluorescent decay rates. As the fiber tip is displaced, the relative intensity contribution of each fluorescent disc changes. Phase-sensitive detection methods are then used to determine the relative contributions of each disc which are proportional to the process variable. The dynamic response of a fluorescent decay-rate sensor is dominated by the fluorescent decay rates of the discs rather than mechanical limits imposed by the diaphragm. Improvements in dynamic response may be possible with the use of other fluorescing material combinations.

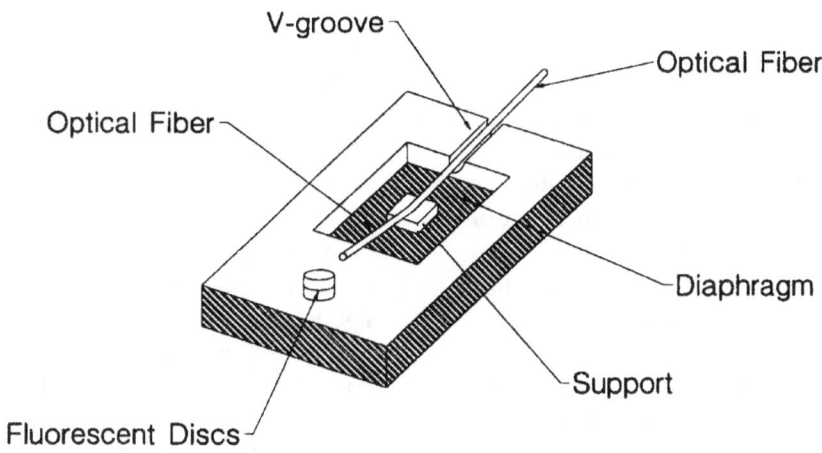

V-groove

Optical Fiber

Optical Fiber

Diaphragm

Support

Fluorescent Discs

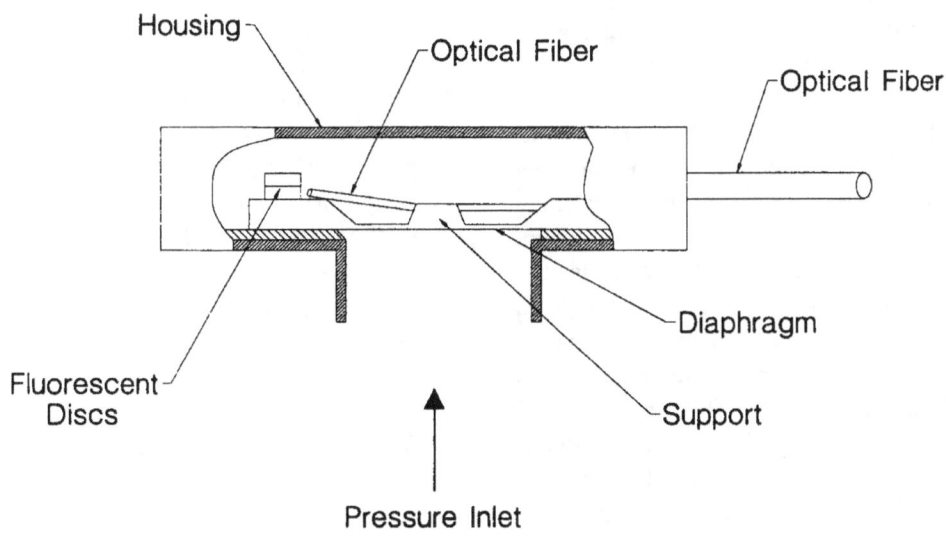

Housing

Optical Fiber

Optical Fiber

Diaphragm

Fluorescent Discs

Support

Pressure Inlet

Figure 5.22   Drawing of a Fluorescent Decay-Time Encoded Pressure Sensor

6. ADVANTAGES AND DISADVANTAGES OF FIBER OPTIC PRESSURE SENSORS

6.1 Advantages

Some of the relative advantages which fiber optic pressure sensors possess over conventional sensors are given in Table 6.1. Although not all fiber optic pressure sensors demonstrate every advantage listed in this table, many demonstrate most of these advantages as discussed below.

Since fiber optic cables are dielectric and the sensing mechanisms used are usually optic and optomechanical, fiber optic pressure sensors are immune to radio frequency interference (RFI) and electromagnetic interference (EMI).[3] This property allows high signal-to-noise ratios and transmission without shielding requirements in electrically noisy environments. As with EMI and RFI, noise emission and cross-talk among fiber optic cables is nonexistent.[26] Optical fibers may be bundled together and several bundles may be run in close proximity.

The dielectric property of fiber optic cables eliminates shock hazards, as well as the need for surge protection and signal isolation.[3] Fiber optic sensors are immune to ground faults and electrical hazards (e.g., ground loops, common mode voltages, and changes in ground potential).[27] Typical materials used in fiber optic cables are chemically inert in most process environments.[1] This property both protects the integrity of the cable and prevents undesirable chemical reaction with or contamination of the process environment.

Transmission loss, or attenuation, is generally much smaller in optical fibers than in the electrical leads of conventional process sensors. With the inherent capability of many fiber optic sensors,[28] several sensors can be used with a single transmission cable. Since some fiber optic sensing modulation techniques are digital in nature, fiber optic sensing is easily made compatible with digital control systems.[29] Many fiber optic transducer designs are extremely small and lightweight.[30] This allows sensor placement in locations previously inaccessible with conventional sensing technologies.

The small size and weight of fiber optic transducers renders them less vulnerable to vibration and shock.[3] Fiber optic pressure sensors offer significantly greater sensitivity[31] and dynamic range[32] than conventional pressure transmitters, and can provide advantages in the following areas: linearity,[33] stability,[34] reliability,[35] and response time.[30]

High temperature environments can introduce diaphragm "creep" and unpredictable and non-repeatable measurements in conventional pressure transmitters. Many fiber optic pressure sensors have been designed so that they are unaffected by elevated temperatures.[32,36] For this reason, as well as tolerance to vibration and shock, fiber optic pressure sensors can be used in adverse environments.[37]

6.2 Disadvantages

Fiber optic sensors have some disadvantages when compared to conventional sensors. These may include fragility of the sensing element and fiber optic cable, poor compatibility with

TABLE 6.1

Advantages of Fiber Optic Pressure Sensors

Optical Isolation Advantages

EMI/RFI Immunity

Noise, Crosstalk, and Ground Loop Immunity

Elimination of Spark and Shock Hazards

Useful in Explosive Environments

Low Signal Attenuation for Remote Measurements

Physical Factors

Small in Size and Mass

Resistant to Harsh Environments

High Temperature Tolerance

Chemically Inert

High Tolerance to Vibration and Shock

Performance Characteristics

High Resolution

High Dynamic Range

Good Linearity

Temperature Compensation or Low Temperature
Sensitivity

Multiplexing Capability

process environments, radiation-induced darkening of the fiber optic cables resulting in increased attenuation, complex and expensive signal processing equipment and, in certain cases, poor performance in static pressure measurements.[10]

Many fiber optic sensors, especially of the intensity-modulated variety, are particularly vulnerable to fluctuations in the transmission characteristics of the fiber optic cables due to environmental and mechanical stressors, as well as variations in the sensitivity of detectors and the efficiency of light sources with temperature and aging. These specific problems are usually eliminated or reduced with the application of compensation techniques (e.g., use of a reference fiber or dual-wavelength measurement), and are not necessarily present with other modulation techniques.[27]

6.3 Intercomparison of Fiber Optic Sensors

In order to highlight the specific advantages and disadvantages of fiber optic sensors, the four different types of fiber optic sensors are compared below.

6.3.1 Intensity-Modulated Sensors

Intensity-modulated sensors, as compared to other fiber optic sensor types, require relatively simple electronics to decode the measurand from the modulated light.[38] This results in a simpler and less expensive device to develop or manufacture.[39] However, some uncompensated intensity-type designs suffer from unintentional attenuation in the cables, as well as variations in the light source and optical detector. If the light intensity is affected by changes other than in the area where the process measurement takes place, then the output of the sensor will be biased by these changes. Intensity-based sensors can be susceptible to the following sources of attenuation: connector alignments or changes in alignment during maintenance, any movement, bending, or vibration of the fiber, radiation-induced darkening, and changes in source intensity or optical detector sensitivity.[10]

The effects mentioned above can all be reduced or alleviated by using a reference fiber or a reference light beam or both.[40] However, intensity-based sensors mostly fall into the extrinsic sensing category, and inherit certain susceptibilities. Extrinsic sensors suffer from the following difficulties which stem from having the light exit from and return to the fiber optic cable: "problems with alignment, vibration, contamination of surfaces, and possible stray light effects."[10] These affects may be compensated to some extent by using a reference light beam, but may lead to degraded sensor performance and sensor failure.

The microbend intensity-type sensor is an intrinsic sensor, and does not inherit the liabilities of extrinsic sensing. In addition, this sensor has been well characterized and demonstrated with various compensation techniques.[38,41] However, microbend sensors have been reported to be susceptible to calibration drift due to aging effects.

6.3.2 Phase Modulated Sensors

Interferometric sensors are less affected than intensity-modulated sensors by irrelevant intensity or phase variations in the non-measurement regions of the fiber optic cables.[10] Due to the extreme accuracy that can be obtained in measuring phase differences, phase-modulated sensors are much more sensitive than intensity-modulated sensors.[39] However, phase-modulated sensors are also generally more expensive.[42]

The sensitivity of interferometric sensors makes them more cross-sensitive to environmental effects, when compared to other fiber optic pressure sensors.[43] Cross-sensitivity has occasionally been demonstrated to limit the ability to measure static pressure,[44] as slowly varying environmental effects can be difficult to compensate for and may induce significant drift.

The Fabry-Perot interferometric sensors offer simpler implementation than other interferometric methods because they do not require a reference fiber. This modulation technique also offers higher sensitivity than other interferometric methods.[19]

6.3.3 Spectrum-Modulated Sensors

Measurement of the modulation of spectral components of light can eliminate most, if not all, of the sensor's sensitivity to intensity fluctuations or attenuation.[27] This inherent compensation is limited, since it is based on the assumption that all losses are wavelength independent. This assumption may not always hold true.

Fabry-Perot etalon based sensors, which enjoy the loss-compensation benefits of wavelength encoding, can be implemented so that detection of the intensity of only two narrow wavelength bands is necessary. These etalon configurations share the same simplicity of implementation as intensity-modulated sensors.[45]

6.3.4 Time or Frequency-Based Encoding Techniques

Intensity fluctuations associated with source efficiency, detector sensitivity and attenuation may be addressed by encoding the measurand using frequency or decay time. Light modulation in this manner can almost entirely eliminate the sensitivity of the transducer to apparent intensity fluctuations.[27]

The force-dependent resonating element quartz sensor is recognized to be extremely insensitive to attenuation and source/detector influences. The optical detector and remote electronics are only required to count the number of pulses in a specified amount of time and are therefore immune to attenuation of the signal.[46] Intensity compensation schemes for other sensing techniques involve comparisons of relative intensities, and are based on assumptions concerning the nature of the fluctuations (e.g., uniform attenuation with respect to wavelength). Frequency-modulation offers the most robust insensitivity to intensity fluctuation mechanisms (e.g., attenuation losses, detector sensitivity, and source efficiency), as intensity variation and fluctuation will not affect the sensor output until the optical power is reduced below the minimum level discernable by the detector.

7. FAILURE MODES

The components of a fiber optic pressure sensing system may be separated into the following subsystems: optical source, transmission medium, transducer, and the optical detector. The failure modes and degradation mechanisms of each of these individual components must be considered in qualification testing and reliability estimation for use of these sensors in safety related instrumentation in nuclear power plants.

Although fiber optic pressure sensing cannot be considered a fully characterized technology from the standpoint of operating experience, the use of fiber optics in the communications field has been well demonstrated. In addition, extensive reliability and failure mode analyses have been performed on those components mutual to both fiber optic sensing and communication systems.[16,47] In addition to stressors common to many other applications, the effects of gamma and neutron radiation fields must be investigated in order for sensors to be installed in nuclear power plants. Radiation effects on fiber optic system components have been investigated but more testing on long-term effects is needed.[47,48]

A detailed assessment of the failure modes of process transducer mechanisms is not included here, as these components are specific to each separate sensor design. For many sensors, especially extrinsic sensors, the main transducer degradation mechanism is misalignment caused either by vibration, shock, or poor assembly and installation. Alternatively, except for those sensors which are considered non-contact (reflective intensity-based and Fabry-Perot etalon wavelength-modulated sensors) and the interferometric sensors, pressure and temperature cycling may result in eventual mechanical breakdown or misalignment of transducer components.

7.1 Failure Modes of Optical Sources

The typical definition of failure for light emitting diodes (LEDs) is a fifty percent reduction in optical power output. A common failure definition for laser diodes is a fifty percent increase in threshold current. This increase in threshold current also results in a decrease in delivered output power for the laser diode.

Both LEDs and laser diodes are subject to either catastrophic failure, where power reduction is abrupt, or gradual failure over a period of time. Gradual failure can be offset by an increase in the current supplied by the drive electronics, but such an increase leads to overheating of the source and eventual catastrophic failure.

The failure modes associated with LEDs are the following: rapid degradation due to the formation of dark line defects (DLDs) and dark spot defects (DSDs), and slow degradation due to thermally driven diffusion of impurities into the device. DLDs and DSDs are caused by material impurities or crystal lattice defects generated in the material during manufacturing. However, the adverse effects produced by these impurities may not develop for many years. DLDs and DSDs in LEDs may be reduced by improving the following manufacturing factors: material selection, device fabrication, and quality control.[47] Slow degradation describes the reduction of output power over time as diffusion of impurities gradually "contaminates" the purity of the active region of the LED. This degradation increases with temperature, and is not a result of manufacturing defects.

The mechanisms of failure in laser diodes include the incubation mode, formations of DSDs, and laser "wear-out." The incubation mode consists of manufacturing defects, and is an early failure mode. This failure mode may be avoided with the application of simple screening and burn-in procedures. DSDs are formed in laser diodes as component degradation results in increased thermal resistance between the heat sink and the laser device. As the temperature builds up in the laser diode, DSDs are formed, decreasing the output power. The formation of DSDs may be minimized by including a protective layer, known as a passivation layer, in the device. [16]

Laser wear-out is attributable to the following conditions: material degradations due to ambient temperature, which may be reduced by decreasing operating temperature; facet oxidation staining caused by photo-oxidation in high humidity environments, which is reduced by the application of a thin layer of a special coating; and crystal lattice defects which lead to the formation of DLDs. Crystal lattice defects may be avoided or reduced by improved material selection and quality control.

Radiation testing has been performed on LEDs and no significant degradation was detected up to gamma radiation doses of 10^5 Grays (Gy), with output power reduction of only five percent at doses of 10^6 Gy.[47] A study of the neutron radiation effect on LEDs demonstrated that neutron fluences up to 3 x 10^{14} n/cm^2 had no effect on output power.[47] Note that the devices chosen for these studies were selected based on their potential radiation hardness.

7.2   Failure Modes of Optical Fibers

Failures in fiber optic cables are manifested in increased signal attenuation or in breakage of the fiber. Attenuation sources are hydrogen migration into the fiber, creation of OH groups, formation of micro-cracks, and gamma and neutron radiation-darkening of the fiber.

Hydrogen may migrate into the cable during temperature cycling in a moist environment, leading to degradation in the physical strength of the jacketing and coating of the fiber cable, as well as changing the optical properties of the core. Hydrogen impurities may also be initially present in the cable after manufacture and may lead to changes in the optical properties of the fiber core. Hydrogen migration may be reduced by selecting cable materials which do not generate hydrogen and cable coatings which are resistant to the ingress of hydrogen.

Fiber cable micro-cracks are generated by bending stresses in the cable. Micro-cracks may be avoided by adhering to bending radius and fiber handling specifications. Also, the proper selection of cable materials can reduce the creation of micro-cracks during thermal cycling due to the differences in thermal coefficients of expansion.

Radiation darkening has been demonstrated in some fiber optic cables to significantly increase attenuation of the light signal.[48] This effect is dynamic, as attenuation is shown to increase with radiation dose, and to decrease with photo-bleaching and annealing. Photo-bleaching and annealing reduce the adverse effects of radiation darkening by the application of optical power and temperature. Selection of certain fiber materials, such as pure silica, can result in fibers that are extremely radiation resistant. Such pure-silica fibers are considered appropriate and suitable for use even in high radiation environments.

Connectors, splices, and couplers which serve to direct and transmit light, are functionally considered to be part of the fiber optic cable for the failure mode analysis. These devices may introduce substantial losses in the fiber optic system. The failure modes of these devices include poor alignment, contamination of surfaces, aging due to handling and re-mating, and vibration. These mechanisms may be alleviated or eliminated with proper handling and alignment procedures.

7.3   Failure Modes of Optical Detectors

The failure modes for optical detectors include increased dark current and electrochemical oxidation. These devices are also sensitive to ionizing radiation as well as intense optical radiation. However, certain optical detectors have been shown to operate reliably in high radiation environments, with negligible reduction in optical responsitivity.[47]

Dark current is the reverse current flow of a reverse-biased diode in the absence of a light signal. The common failure definition or condition for these devices is an order of magnitude increase in dark current. Dark current occurs with all diodes, and is dependent on operating temperature (approximately doubles for each 10 degrees Celsius increase in temperature). This effect may be reduced by lowering the ambient temperature of the electronics.

High relative humidity has been found to cause electrochemical oxidation in optical detectors. This leads to electrical short circuits and possibly catastrophic failure. This failure mode may be avoided by hermetically sealing electronic components that operate in high humidity environments.

8. QUALIFICATION CRITERIA FOR USE IN NUCLEAR POWER PLANTS

This chapter presents a review of the key points in two of the existing IEEE standards for qualification testing of safety-related equipment such as pressure transmitters in nuclear power plants. The purpose of this review is to provide the reader with an idea of the level of effort that may be involved in qualifying a fiber optic pressure sensor for use in nuclear safety-related applications. Of course, special test procedures based on the general guidelines in IEEE standards will probably be developed by fiber optic sensor manufacturers when the sensors are ready to be qualified for use in nuclear power plants. The current procedures for qualification testing of conventional pressure transmitters may not suffice due to some major differences between the principle of operation of conventional pressure sensors and fiber optic pressure sensors.

8.1 IEEE Standards

Safety-related electrical equipment is referred to as "Class 1E" equipment as defined in the Institute of Electrical and Electronics Engineers (IEEE) Standard 323-1974, commonly referred to as IEEE 323-1974. This standard describes the basic requirements for qualifying Class 1E equipment. A quote from this standard is given below to identify the goal of the qualification tests.

"These qualification requirements, when met, will confirm the adequacy of the equipment design under normal, abnormal, design basis event, post

design basis event, and containment test conditions for the performance of Class 1E functions."[49]

The IEEE 323-1974 includes seismic qualification requirements which are described in IEEE 344-1975. The IEEE 344-1975 is used as a supplement to IEEE 323-1974.

The scope of IEEE 344-1975 is quoted below to show the intent of the seismic qualification tests.

"These recommended practices provide direction for establishing procedures that will yield data which verify that the Class 1E equipment can meet its performance requirements during and following one SSE (safe shutdown earthquake) preceded by a number of OBEs (operating basis earthquakes)."[50]

Qualification, according to the IEEE standards, can be accomplished in several ways: type testing, operating experience, analysis or a combination of the three. Type testing, which involves subjecting the equipment to normal and abnormal conditions, would probably be the first step in qualification testing of a new pressure sensor. Qualification by analysis would involve the generation of mathematical models to simulate the effects of potential conditions that the equipment may experience in a plant. This typically implies that data, short term and long term, would be used to corroborate the assumptions made in the mathematical models. For fiber optic pressure sensors, qualification by analysis could present problems because sufficient operating data is not yet available from industrial applications of these sensors. For the same reason, qualification from operating experience may not be feasible for fiber optic pressure sensors. Therefore, qualification by type testing as described below would probably be the most suitable method for fiber optic pressure sensors.

8.2 Qualification Testing

Type testing methods for qualification of Class 1E equipment are presented in IEEE 323-1974. Table 8.1 provides a listing of the equipment characteristics that are monitored during these tests. The qualification tests are performed at conditions which meet or exceed the conditions that the equipment will be exposed to during its useful life including both normal and abnormal operations.

Categories II and V in Table 8.1 deal with the electrical characteristics of the equipment. As discussed earlier in this report, fiber optic pressure sensors are electrically isolated in the field and their electrical components will usually be located in a mild environment such as the control room area. Therefore, in writing qualification test procedures for fiber optic pressure sensors, these categories may not apply in the same way that they do for conventional pressure sensors. For example, Category V discusses monitoring insulation resistance, voltage and current, while for fiber optic sensors, the qualification tests should involve monitoring such factors as optical frequency, attenuation, and distortion.

The sequence in which the qualification tests are performed is also specified in the IEEE 323-1974 standard. A typical test sequence is illustrated in Figure 8.1. The Design Basis Event (DBE) testing may follow the test profile shown in Figure 8.2. The dotted line shows the conditions that the equipment will be subjected to. The solid line shows the actual test parameters based on postulated plant conditions plus performance margins which are added to account for variations in the production of the equipment and possible errors in defining its performance requirements.

8.3 Aging Tests

According to IEEE 323-1974, aging tests can take several forms. If previous aging data for specific components exist, they are used where applicable. Alternatively, accelerated aging tests may be used including the effects of heat, vibration, radiation, etc. Thermal aging, for example, is performed according to the Arrhenius equation. This equation defines the temperature and time period necessary to thermally age the equipment so that it can be tested to verify that it will operate properly throughout its useful life.

The component of a fiber optic sensor that may be susceptible to aging degradation is the fiber optic cable. As discussed in Chapter 6, one of the major drawbacks of fiber optic sensors for nuclear power plants is the effect of radiation on fiber optic cables. Although advances in the manufacturing of these cables have reduced this weakness, extensive testing may be necessary to ensure that fiber optic sensors will operate properly in a nuclear power plant for a sufficiently long period of time.

TABLE 8.1

**Equipment Characteristics to be Monitored
During IEEE 323-1974 Class 1E Qualification Testing**

Category	Title	Description
I	Environment	Temperature, pressure, moisture content, gas composition, vibration, and time
II	Input Electrical Characteristics	Frequency, current, voltage, power to the equipment, and time duration of the input
III	Fluid Characteristics	Concentration of chemical constituents in fluid injected into the test chamber plus the flow rate and spray disposition and temperature of such fluids
IV	Radiological Features	Nuclear radiation data including energy type, energy level, exposure rate, and integrated dose
V	Electrical Characteristics	Insulation resistance of electrical components; voltage, current and power output; response time; frequency characteristics and simulated load
VI	Mechanical Characteristics	Thrust, torque, time, and load profile
VII	Auxiliary Function Measurements	Function measurements related to Class 1E equipments which are included in the equipment but not necessary for its own operation; that is items which are required to provide a signal to control other Class 1E equipment

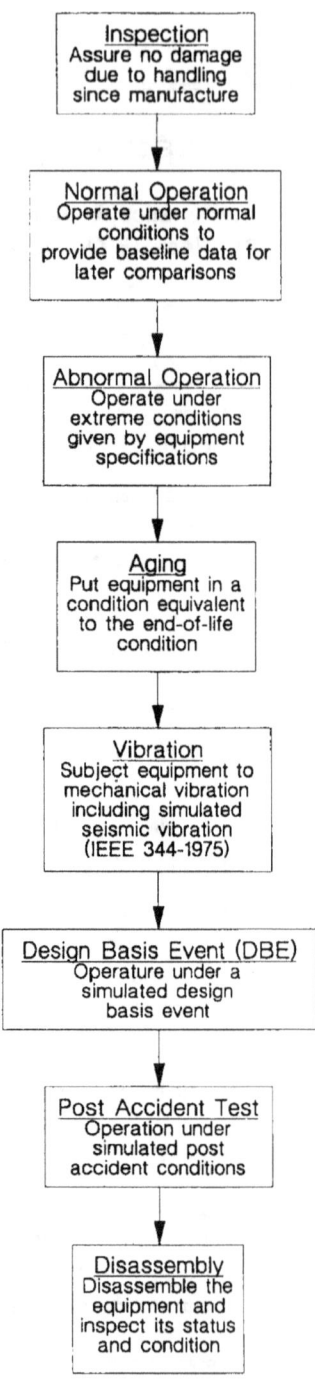

**Figure 8.1   Flow Chart Showing the Qualification
Test Sequence for Class 1E Equipment**

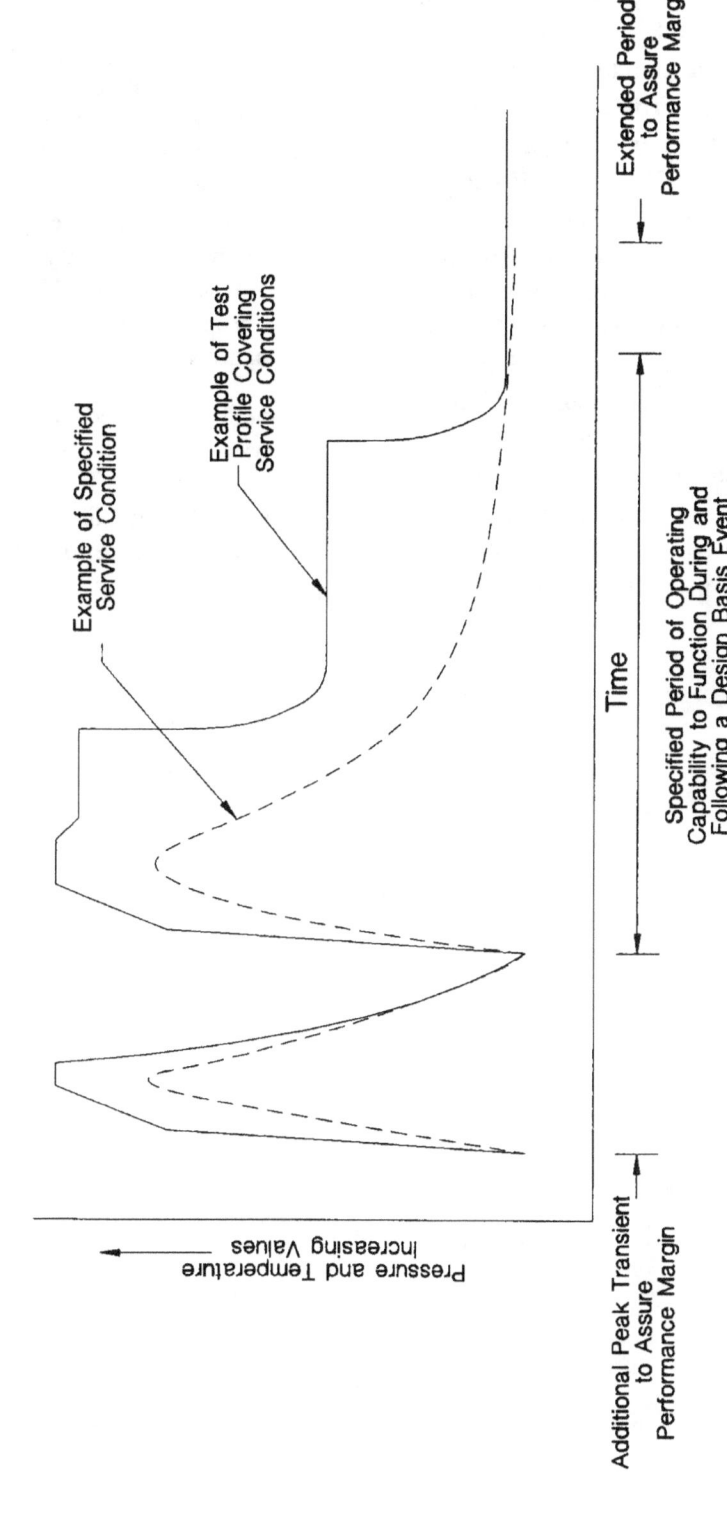

Figure 8.2 Design Basis Event Testing Profile for Qualification Testing of Class 1E Equipment

Although the electronic components in a fiber optic sensor would normally be located in controlled areas and subject to mild environments, qualification testing may still be necessary for the electronics. The average temperature in containment for nuclear power plants may approach 120 degrees F and the gamma radiation dose levels may be on the order of 30,000 Gy over a sixty five year period. For the control room environment, the temperature is typically between 65 and 70 degrees F while the dose level over sixty years would typically be less than 10 Gy.[47] However, radiation is still a potential source of failure in fiber optic electronics, especially in the light sources and optical detectors as outlined in Chapter 7. The qualification testing of the electronics, although less intense, would require aging tests similar to that for the cables and the transmitter in the field.

Overall, qualifying a fiber optic sensor for nuclear safety-related applications would probably be a major task. In fact, discussions with manufacturers of nuclear plant pressure sensors conducted as a part of this project have indicated that the qualification issue is one of the reasons why these manufacturers have not seriously considered developing fiber optic sensors for nuclear safety-related applications.

It should be pointed out that qualification of a sensor as Class 1E equipment is more involved than what has been described in this chapter. The qualification question will be addressed in more detail in the Phase II project.

9. INDUSTRY SURVEY RESULTS

A critical aspect of this research project was the accumulation of knowledge from the individuals and organizations involved in fiber optic sensing technologies. This was accomplished by an informal survey of the fiber optic manufacturers and interviews with the experts in this field. Along with the information obtained from the literature reviewed, the manufacturers survey and the interviews provided comprehensive information on the state-of-the-art in fiber optic pressure sensing as well as the advantages, disadvantages and failure modes of fiber optic pressure sensors. Discussions and site visits with manufacturers of fiber optic sensors, as well as various customers, researchers and acknowledged industry experts, provided an objective assessment of the current interest in and potential of fiber optic pressure sensors for use in the nuclear industry.

Appendix A contains a listing of the relevant literature obtained and reviewed during this research effort. Included in this listing are papers, journal and magazine articles, books, and other publications pertaining to fiber optic sensing as well as product bulletins from manufacturers of fiber optic sensors.

9.1   Description of Survey

Approximately one hundred individuals and organizations in the United States, Canada and European nations were contacted during the Phase I effort. These contacts represented a wide variety of expertise in fiber optic sensing technology. Figure 9.1 gives a breakdown of the contacts made during the project by identifying their involvement in fiber optic sensing. The majority of the organizations contacted were manufacturers of fiber optic sensors and related components. However, this breakdown is somewhat arbitrary in that many of the individuals surveyed could fall into several categories. For instance, many of the fiber optic sensor manufacturer contacts were also authors of various publications. This also affects the "Researcher" category because many of the manufacturers also perform extensive research on fiber optic technologies. However, for simplicity, the survey contacts were placed into a category that was most related to their overall involvement in this technology.

Two questionnaires were used in performing the survey described in this chapter; one for the survey of manufacturers and another for interviews of industry experts, scientists, authors, etc. These questionnaires are included in Appendix B. The questionnaires were mailed to the contacts but only a few written response were received. Subsequently, telephone contacts were made and the questionnaires were completed over the telephone. Note that this was an informal survey with the purpose of determining the state of fiber optic pressure sensing technologies. A scientific survey will be designed and performed in Phase II.

One of the most important questions asked during the survey involved the potential of fiber optic pressure sensors for safety-related measurements in nuclear power plants. Although a wide variety of responses were given to this question, the overall response was positive. Most manufacturers and industry experts felt that the necessary technology is currently available to design a fiber optic pressure sensor that can meet or exceed the nuclear industry design and qualification criteria. However, because of the current lack

Industry Survey Breakdown

JPF218A-01A

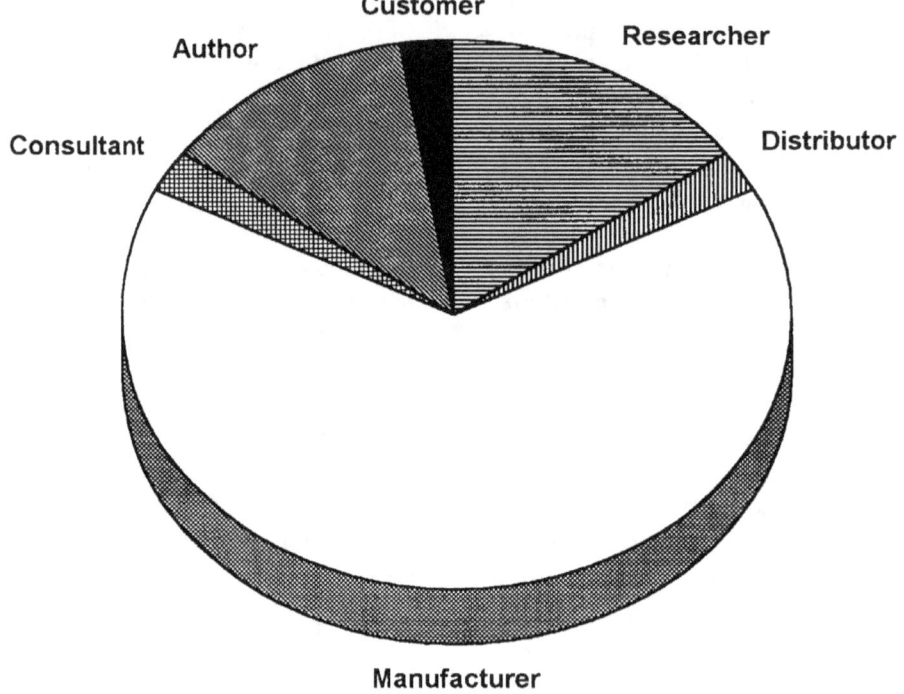

Figure 9.1   Breakdown of the Contacts Made During the Industry Survey

of research and testing into such items as long term radiation effects, this may not be possible in the near future.

9.2   Manufacturers Survey

As mentioned above, most of the organizations contacted were manufacturers of fiber optic sensors and fiber optic sensing system components. These manufacturers were identified through the literature review, searches of manufacturing indices, as well as through information obtained from various individuals contacted during the project. Over sixty companies were contacted in order to establish their involvement in fiber optic sensing, as well as to determine their interest in supplying products to the nuclear industry.

Figure 9.2 gives a breakdown of the involvement of the manufacturers surveyed in the development of fiber optic pressure sensors. As shown in this figure, only a small number (five) of manufacturers actually stock fiber optic pressure sensors. However, most of these sensors are intended for specific industrial uses such as in the chemical, medical, automotive and aerospace industries. Therefore, intercomparisons of the sensors manufactured by these companies was not possible.

Custom manufacturers were defined as those organizations that only design and fabricate fiber optic pressure sensors per customer requirements on an as-needed basis. These companies also typically manufacture fiber optic sensors for measuring other process variables. The manufacturers listed under the "R&D Phase" category in Figure 9.2 are companies that are currently developing a fiber optic pressure sensor for commercialization. However, as for the manufacturers that stock sensors, these companies are typically targeting a niche market.

The three remaining categories shown in Figure 9.2 are companies that manufactured fiber optic pressure sensors at one time but are no longer involved in the technology, companies that went out of business, and companies that only manufacture fiber optic sensors that measure process variables other than pressure. Not shown in Figure 9.2 are companies that manufacture fiber optic system components only. These components include optical sources and detectors, as well as fiber optic cables. Information from these contacts was very helpful in establishing the characterizations and failure modes of the components that make up a fiber optic pressure sensing system.

Chapters 5 and 6 discuss the different types of fiber optic pressure sensors as well as their advantages, disadvantages and failure modes. Although a variety of fiber optic sensing methodologies were described in these chapters, four distinct categories were established. Figure 9.3 shows the results of the manufacturers survey in terms of the popularity of these four sensing techniques. The main reasons given for the popularity of the intensity modulation technique over the other three was its simplicity, ruggedness and low cost.

An evaluation of the popularity of different fiber optic sensing measurands was also performed as shown in Figure 9.4. As seen in this figure, temperature is the most popular measurand although a significant number fall under the "Other" category. This category includes the measurands listed in Table 9.1. The variety of measurands is consistent with the conclusion that most fiber optic sensors have been developed for highly specialized applications. Note that the popularity of pressure may be somewhat biased by the fact that the overall intent of the survey was to evaluate the fiber optic pressure sensing industry. Therefore, manufacturers of other types of fiber optic sensors may have been overlooked.

Manufacturers Survey Breakdown

JPF219A-04A

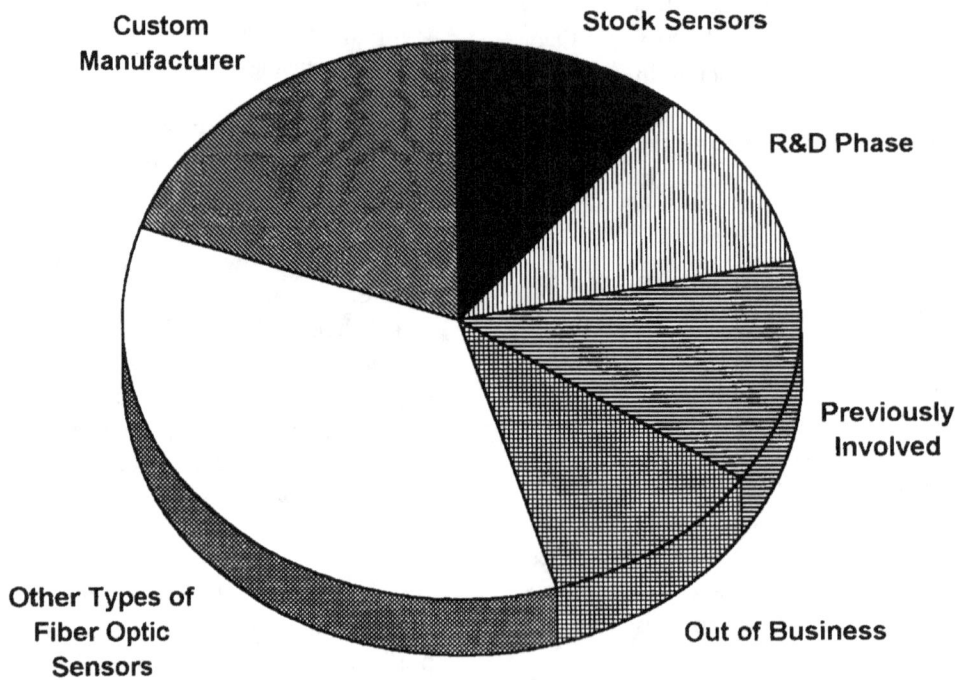

Figure 9.2   Involvement of the Manufacturers
Surveyed in Fiber Optic Pressure Sensing

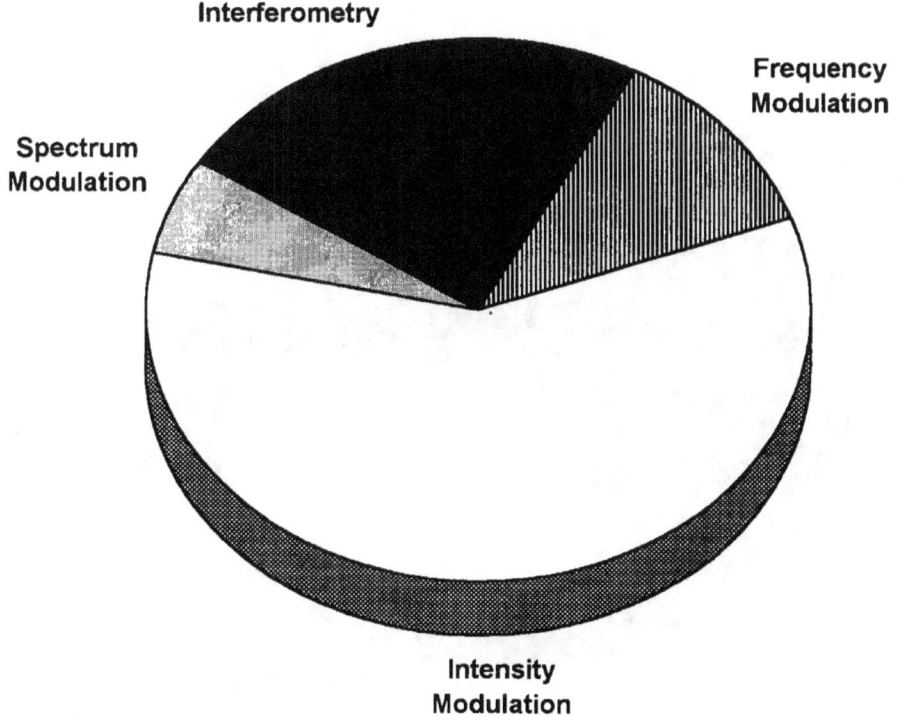

Fiber Optic Pressure Sensing Techniques

JPF218A-02A

Interferometry

Frequency Modulation

Spectrum Modulation

Intensity Modulation

Figure 9.3   Popularity of the Four Main Fiber Optic Pressure Sensing Techniques

**Fiber Optic
Sensing Measurands**

JPF219A-03A

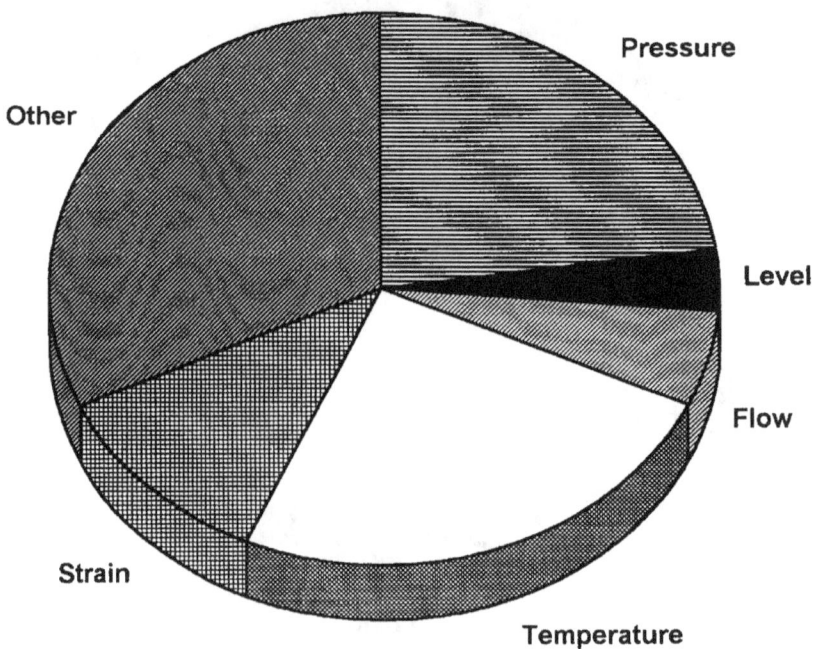

Figure 9.4   Distribution of Measurands for the Fiber Optic
Sensors Identified in the Manufacturers Survey

TABLE 9.1

Other Fiber Optic Sensing Measurands Identified Through the Industry Survey

Air/Fuel Ratio
Bending
Corrosion
Electrical Current
Electromagnetic Fields
Rotation Rate
Refractive Index
Sound
Torsion/Torque
Vibration
Viscosity
Void Fraction

An important part of the manufacturers survey was to establish the current interest of fiber optic pressure sensor manufacturers in entering the nuclear market. Although most were interested, all of these manufacturers had some reservations. Their reservations were mostly due to the design and qualification criteria, as well as skepticism concerning the interest of the nuclear industry in procuring such technologies. Most could not see an economic benefit for meeting the nuclear utility industry's requirements at this time. Also, note that most of the fiber optic pressure sensor manufacturers surveyed were companies with no previous involvement in the nuclear industry. The manufacturers that currently supply conventional pressure sensors to the nuclear industry are only moderately interested in developing fiber optic pressure sensors. A few have performed some limited investigations into this technology for nuclear power plants, but none are vigorously pursuing this issue.

9.3   Babcock & Wilcox Site Visit

Babcock & Wilcox (B&W) was identified as a custom manufacturer of fiber optic pressure sensors. This company has also manufactured fiber optic sensors that measure flow, temperature, strain, displacement and void fraction. Their fiber optic pressure sensors have employed both the microbend and Fabry-Perot interferometry techniques.

Two AMS engineers visited the B&W Applied Measurements Section facilities in Alliance, Ohio in October 1994. They toured the facilities and spoke with individuals involved in fiber optic sensing including Dr. John W. Berthold. Dr. Berthold is an expert in fiber optic sensing technologies and has published many papers on the development and applications of these technologies.

Figure 9.5 is a picture of a microbend fiber optic pressure sensing system developed by B&W. This is an absolute pressure measurement system which can potentially handle pressures up to 2,750 PSI and temperatures up to 300 degrees F. This system also offers temperature correction to provide greater accuracy.

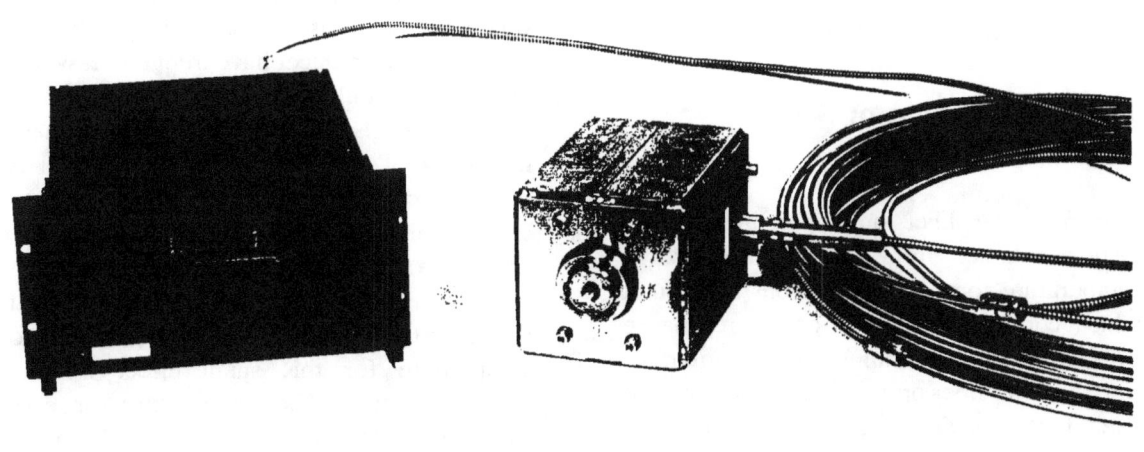

**Figure 9.5   Microbend Fiber Optic Pressure
Sensor Developed by Babcock & Wilcox**

10. LABORATORY TESTING OF A FIBER OPTIC SENSOR

In an attempt to enhance the information compiled in the previous chapters on fiber optic sensing technologies, a fiber optic pressure sensor was obtained from Paroscientific, Incorporated of Redmond, Washington, for laboratory testing. This testing included measurements of the static and dynamic performance of the sensor.

10.1 Description of Sensor

The particular fiber optic pressure sensing system obtained from Paroscientific was a temperature-compensated demonstration unit which consists of a quartz crystal resonance pressure transducer, an electronic interface unit, and two fiber optic cables. The sensor is used for barometric pressure measurements between 11.5 and 16 PSIA (PSI absolute). Paroscientific also manufactures fiber optic sensors which measure pressures up to 40,000 PSIA.

An interface unit supplies the optical power to the transducer through one of the fiber optic cables. The transducer then modulates the light according to the measured pressure and returns the modulated light to the interface unit through the second fiber optic cable. A drawing of the Paroscientific system showing the setup and the dimensions of the transducer and the interface unit is given in Figure 10.1 and a picture of these two components is given in Figure 10.2.

10.1.1 Principle of Operation

The Paroscientific fiber optic pressure transducer tested in the laboratory uses quartz crystal resonance technologies to modulate the frequency of the light signal according to both the sensed pressure and the temperature inside the transducer. This is accomplished by using two quartz crystals, one for pressure and one for internal temperature, which resonate at different frequencies depending on the values of the measurands. The light beams that exit the transducer have frequencies which are dependent upon the resonating frequencies of the crystals. The resonant frequency of the pressure sensing crystal is primarily dependent on the sensed pressure but is somewhat affected by temperature. Therefore, the internal temperature sensing crystal, which has an optical frequency output dependent on temperature alone, is used to compensate for this small effect on the pressure crystal. The temperature range of this particular transducer is -65 to 225 degrees Fahrenheit although similar transducer designs can go up to approximately 250 degrees Fahrenheit.

The major advantages of the frequency modulation techniques for fiber optic pressure sensing are high accuracy, repeatability, low hysteresis and power consumption, along with long term stability.[51] This particular sensor offers a 0.01% accuracy and its stability is comparable to timepieces which use crystal oscillators. However, the quartz crystals themselves are very sensitive to shock, vibration, and other stresses. The results of overstressing the crystals are typically catastrophic in terms of the performance of the sensor. In order to alleviate the potential for such problems, the crystals are mounted on mechanical isolation systems and mounting pads. Balance weights are also included to reduce the sensitivity of the sensor to

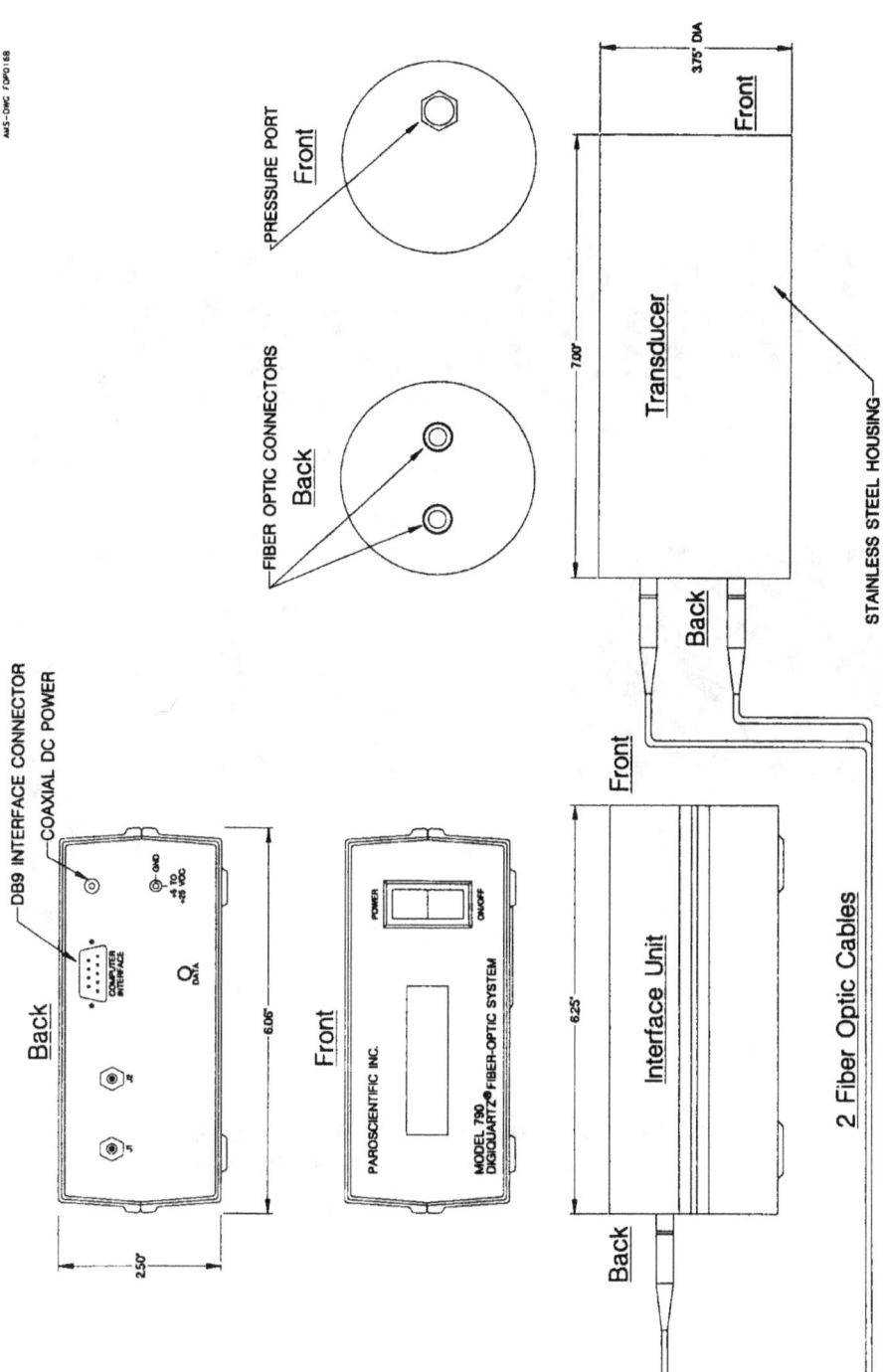

Figure 10.1   Drawing of the Paroscientific Fiber Optic Sensing System

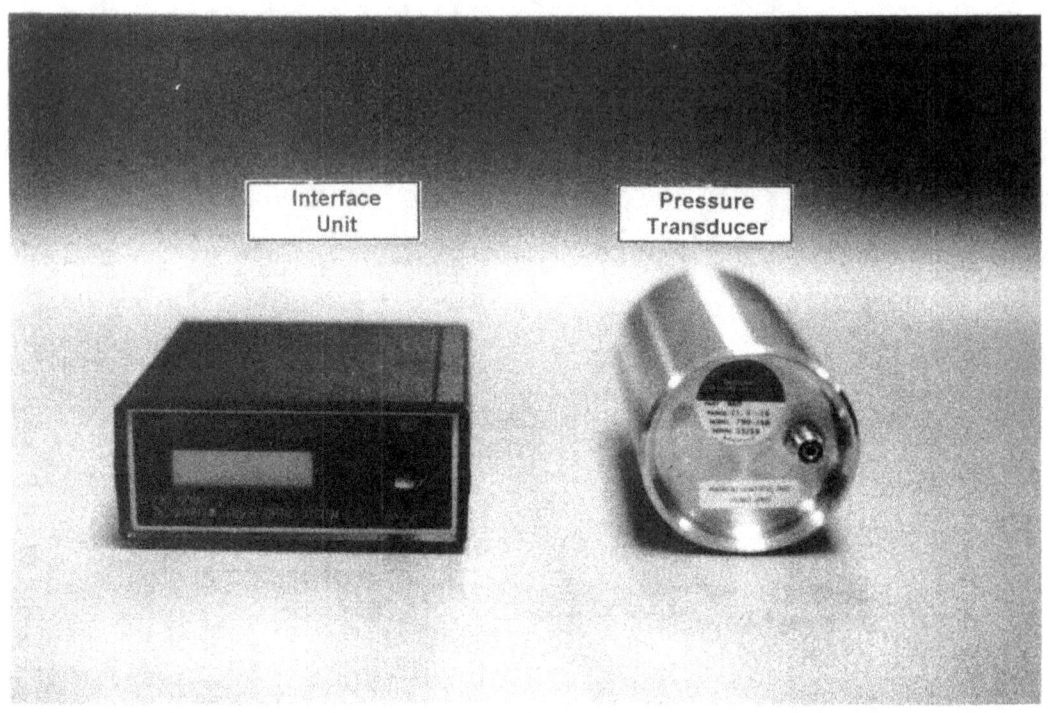

**Figure 10.2   Picture of the Paroscientific Fiber Optic
Pressure Transducer and Interface Unit**

acceleration, shock and vibration. The applied pressure is transferred to the quartz crystal through a bellows configuration in this particular sensor, although a Bourdon tube configuration is used in transducers which measure higher pressures.

Figure 10.3 shows the internals of the Paroscientific pressure transducer. Note that the actual sizes of the components are not accurately portrayed in this figure. The sensing elements are very small compared to the dimensions of the transmitter housing. This transducer is large because it was designed for use in explosive environments. Therefore, a neoprene boot is sandwiched between two cylindrical stainless steel housings to protect the internals of the transducer (Figure 10.4).

10.1.2 Fiber Optic Cables

The light beams are sent to and from the pressure transducer through fiber optic cables which may be up to 500 meters long. The fiber optic cables for the Paroscientific sensor are supplied by the SpecTran Specialty Optics Company. They are multimode step index fibers with a high numerical aperture. Their high numerical aperture make them very tolerant to improper connections as well as macrobending. It also allows a greater amount of the light source to enter the cable thereby increasing optical power and allowing the use of less expensive light sources and optical detectors. Their ruggedness and low cost make them ideal for fiber optic instrumentation. A drawing of the SpecTran fiber optic cable is given in Figure 10.5.

10.1.3 Interface Unit

The electronic interface unit, which can be located remotely from the pressure transducer, supplies light to the transducer and interprets the modulated light signals from it. A digital display is provided with the interface unit to provide local pressure readings. A computer can be linked to the interface unit through a standard RS-232 communications interface. This allows the user to control the interface unit display, receive pressure and temperature measurements with the desired format and resolution, and control other features of the sensing system. The communications protocol used with this system is such that up to 98 different transmitters can be attached to a single RS-232 port and controlled by a single computer.[52]

Figure 10.6 is a block diagram of the entire sensing system that illustrates how the individual components of the interface unit function. The optical receiver/demodulator converts the incoming modulated light signals to electrical pulses. A digital counter, which is driven by a high frequency digital clock, is used to count the number of pulses that occur in a specific amount of time. The time between pulse counts is called the period and is inversely proportional to the frequency of the incoming light signal. A microprocessor is used to control the counter which is multiplexed to allow it to measure the period of both the temperature and pressure signals. The period measurements are then processed by the output circuitry according to the coefficients stored in the memory of the interface unit. This memory is in the form of an Erasable and Programmable Read-Only Memory (EPROM) that contains unalterable information as well as an Electrically Erasable and Programmable Read-Only Memory (EEPROM) that contains measurement conversion characteristics, display modes, and other parameters, which can be changed by the host computer.

10.2 Calculation of Pressure and Temperature

The period measurements from the counter are converted to both temperature and

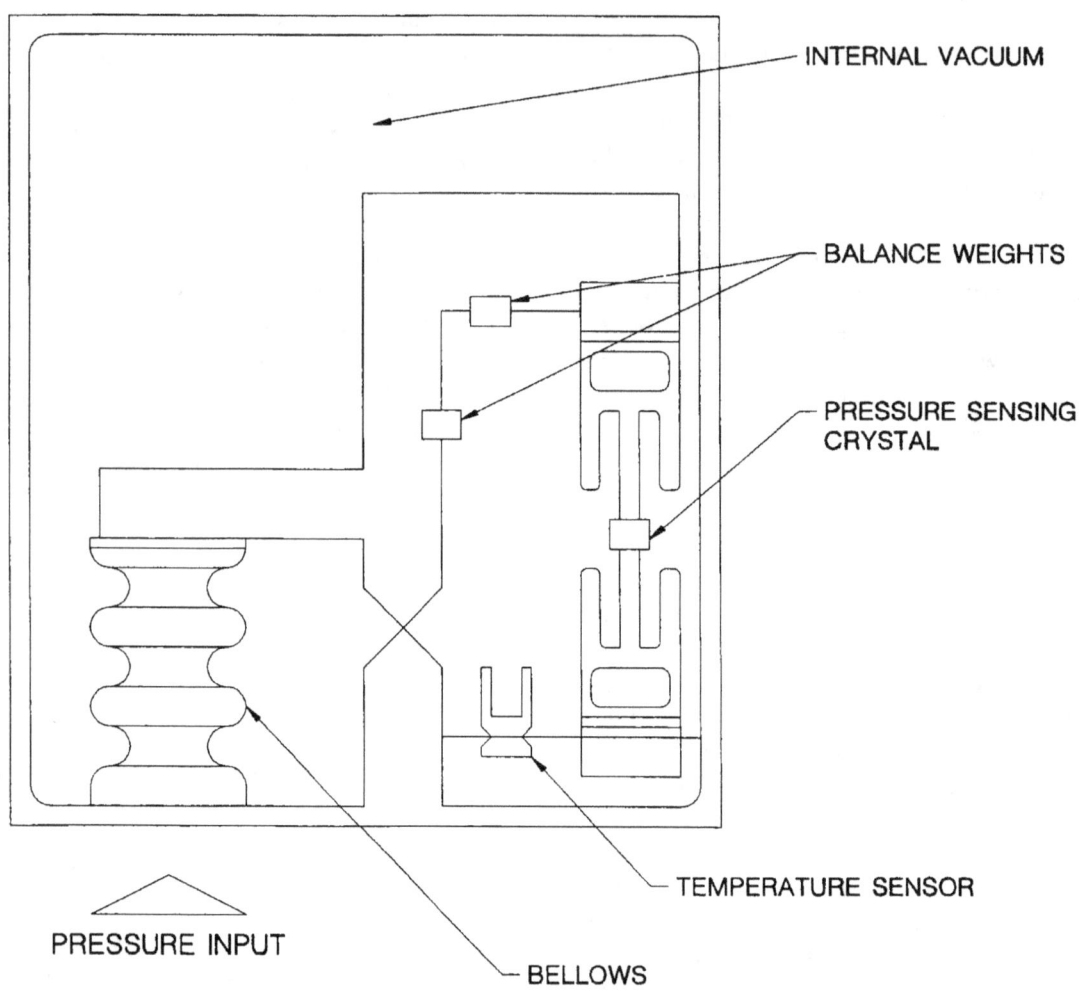

INTERNAL VACUUM

BALANCE WEIGHTS

PRESSURE SENSING
CRYSTAL

TEMPERATURE SENSOR

PRESSURE INPUT

BELLOWS

Figure 10.3   Drawing of the Internal Components of the
Paroscientific Fiber Optic Pressure Transducer

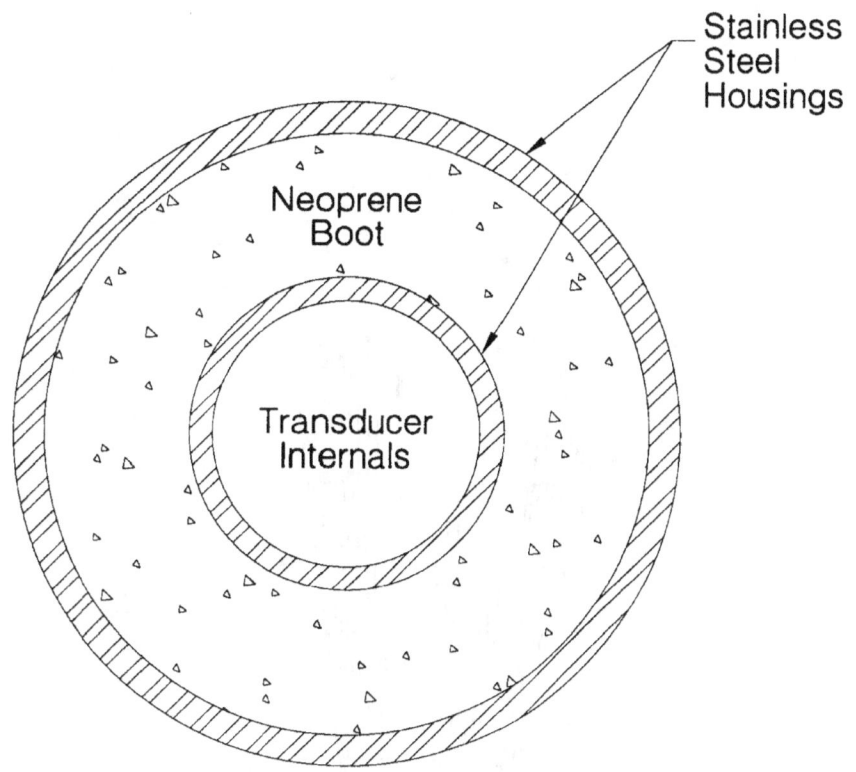

Figure 10.4 **Illustration of the Protective Housings Placed Around the Internals of the Fiber Optic Pressure Transducer**

Pure Silica Core

Bonded Hard
Polymer Cladding

Tefzel® Buffer

Kevlar® Braid

Polyurethane
Outer Jacket

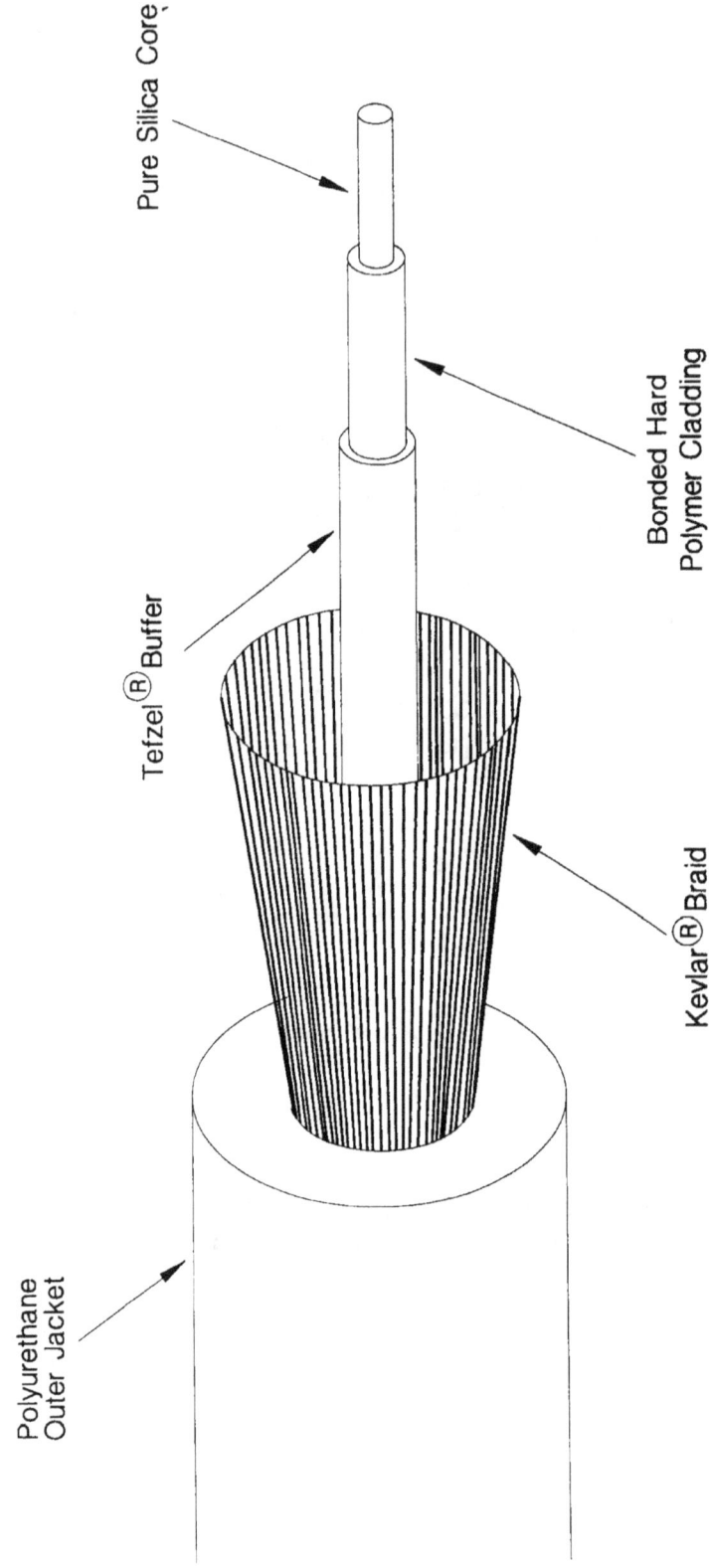

**Figure 10.5   Drawing of the Fiber Optic Cable Used
with the Paroscientific Pressure Sensor**

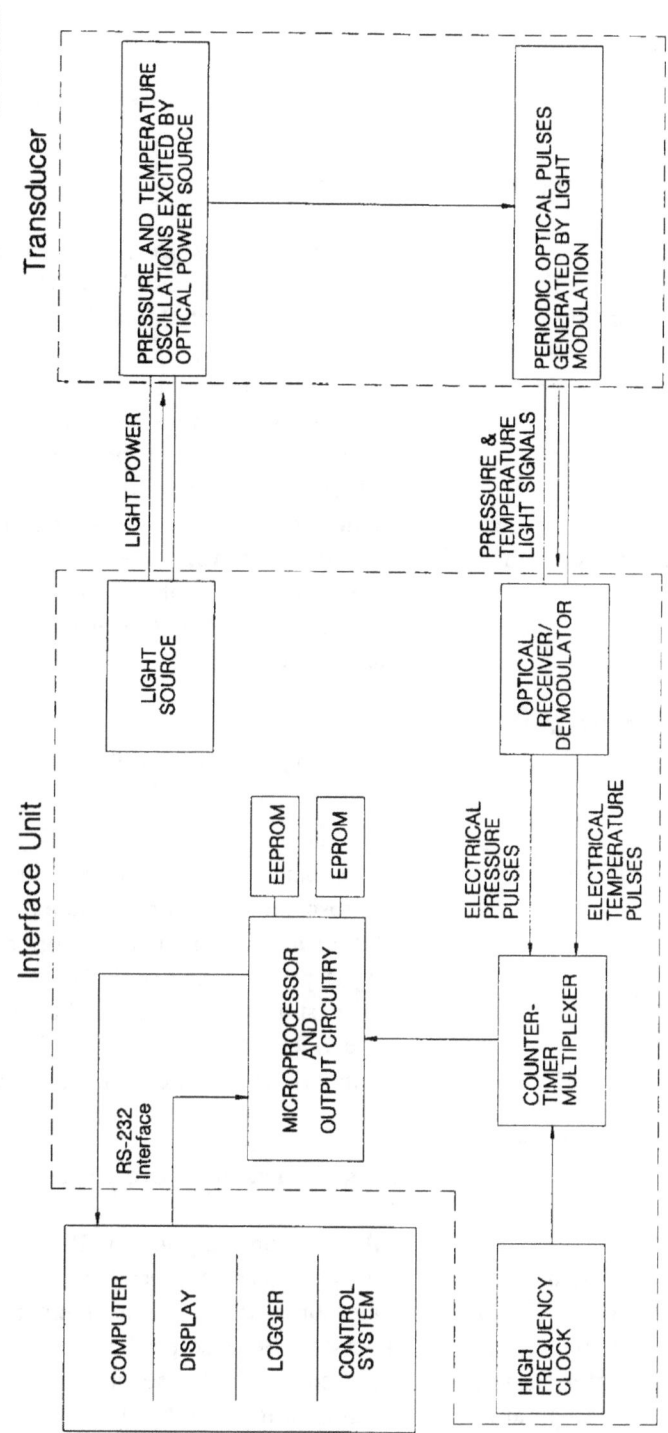

Figure 10.6   Block Diagram of the Paroscientific Pressure Sensor Showing the Components of the Transducer and the Interface Unit

pressure based on the calibration coefficients stored in the EEPROM. The calculation of the transducer's internal temperature is performed as follows:

$$\text{Temperature (°C)} = Y_1U + Y_2U^2 + Y_3U^3$$

where

$$U = \text{Temperature Period} - U_0 \text{ microseconds}$$

and

$$Y_n \text{ and } U_0 = \text{Calibration Coefficients}$$

The temperature-compensated pressure can be calculated from both the pressure and temperature periods as follows:

$$P = C(1 - T_0^2/Tau^2)[1-D(1-T_0^2/Tau^2)]$$

where

$$C = C_1 + C_2U + C_3U^2$$

$$D = D_1 + D_2U$$

$$T_0 = T_1 + T_2U + T_3U^2 + T_4U^3 + T_5U^4$$

and

$$Tau = \text{Pressure Period in Microseconds}$$
$$C_n, D_n \text{ and } T_n = \text{Calibration Coefficients}$$

The relationship between the temperature inside the transducer and the temperature-compensated pressure (P) is shown in Figure 10.7. In this figure, the pressure period measured by the interface unit is held constant to reveal the temperature compensation characteristics of the device. The actual pressure output displayed by the

interface unit and sent to the host computer is calculated as follows:

$$P_{output} = PM[(UM \cdot P) + PA]$$

where

PM = Pressure Multiplier (Span)
PA = Pressure Adder (Zero)
UM = Units Multiplier

The units multiplier term can be used to display the measured pressure in units other than PSIA while the pressure multiplier and adder terms allow the user to alter the calibration of the sensor without disturbing the original calibration coefficients. The original coefficients are calculated and stored in the interface unit by the manufacturer before shipment.

10.3 Laboratory Testing

The laboratory testing performed with the Paroscientific sensor was aimed at providing a better understanding of the operation of fiber optic pressure sensors. The testing involved both static and dynamic characterizations of the fiber optic sensor and demonstrated its temperature compensating abilities.

10.3.1 Temperature Cycling

As mentioned earlier, the Paroscientific sensor uses a second crystal in the design that measures the internal temperature of the device. Because of the protective enclosures placed around the sensing elements, there is a significant lag between the external temperature and the temperature seen by the sensing elements. This temperature lag is

Temperature-Compensated Pressure

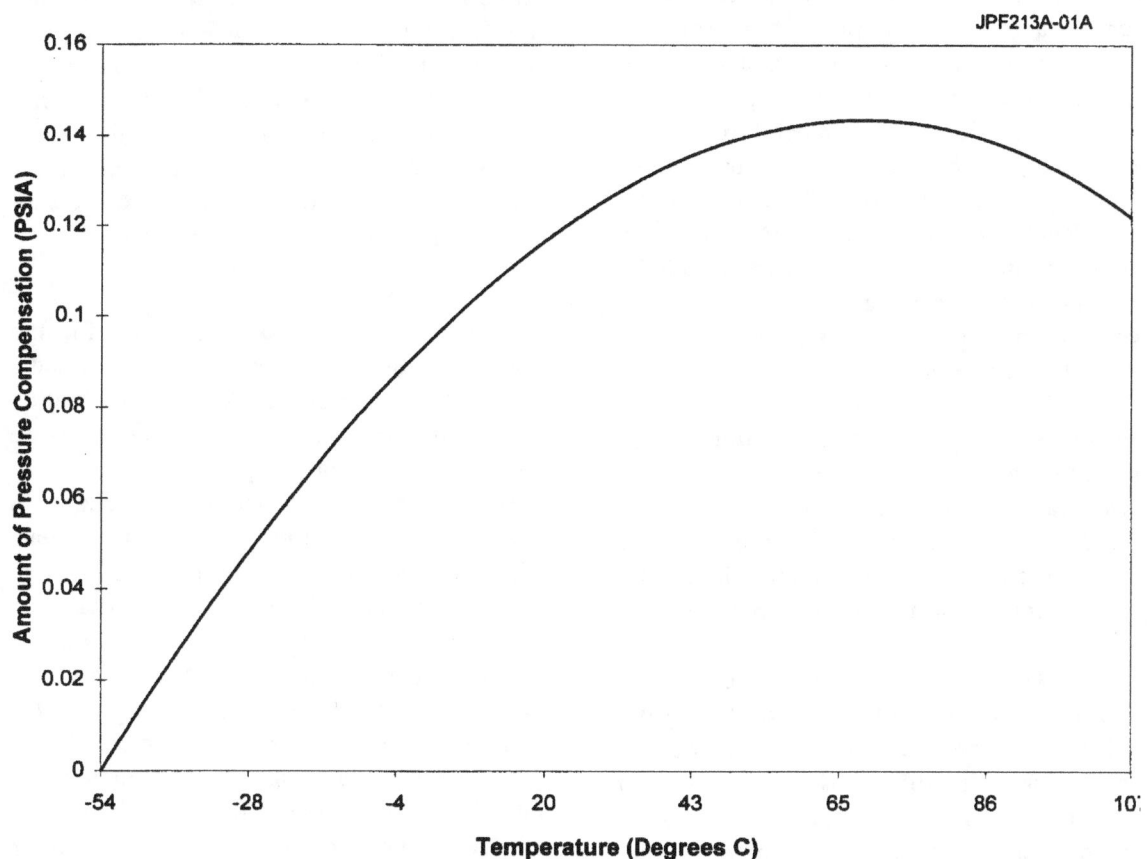

Figure 10.7   Relationship Between the Transducer Internal Temperature and the
Temperature-Compensated Pressure Output for the Paroscientific Sensor

shown in Figure 10.8. An environmental chamber that had been used in previous instrumentation aging research projects was utilized for the temperature cycling tests. The Paroscientific sensor, along with a conventional pressure sensor, is shown in the environmental chamber in Figure 10.9.

In order to demonstrate the temperature compensation capabilities of the Paroscientific sensor, a conventional pressure sensor was placed outside the environmental chamber as a reference while the fiber optic sensor remained inside. The temperature in the environmental chamber was cycled from room temperature, approximately 70 degrees F, to 150 degrees F. The temperature was then held constant for a period of approximately 13 hours before the heating was stopped and the environmental chamber was allowed to cool down to room temperature.

Figure 10.10(a) shows the temperature of the environmental chamber versus the internal temperature of the fiber optic sensor for one test run. Figure 10.10(b) shows the output of the conventional sensor versus both the uncompensated and temperature-compensated pressure signals coming from the fiber optic sensor. The uncompensated pressure signal, as shown in this figure, follows the outline of the reference sensor output but shows an increasing deviation as the internal temperature increases. The temperature-compensated pressure output, however, more closely follows the outline of the conventional sensor. Note that the intent of this test was to demonstrate the temperature compensation abilities of the fiber optic sensor and not to determine its accuracy. Figure 10.10(c) illustrates the temperature compensation more clearly by showing the deviation of the temperature-compensated measurement from the uncompensated output of the fiber optic sensor.

10.3.2 Vibration Testing

One of the major concerns with the Paroscientific sensor was the potential damage to the sensing crystals due to vibration and other mechanical stressors. As mentioned above, a mechanical isolation system including mounting pads and balance weights is employed in this sensor to reduce the adverse effects of such stressors and protect the crystals. Because the Paroscientific sensor was a demonstration unit on loan for this research project, destructive vibration testing could not be performed. Therefore, the intent of the vibration testing was to demonstrate the ability of the fiber optic sensor to measure pressure accurately despite induced vibration.

Figure 10.11 shows the test setup for the vibration testing. The fiber optic sensor and a conventional sensor were fixed to a vibration beam and exposed to the same pressure source. The beam was pinned at one end to restrict movement to only one direction. A motor was fixed to the other end of the beam to induce vibration. Various size weights were tied to the shaft of the motor to produce an imbalance and thereby increase the vibration of the beam. Figure 10.12 shows the results for one set of vibration tests by comparing the output of the Paroscientific sensor to that of the conventional sensor. The motor was periodically turned on and off, as shown in this figure, to illustrate that the induced vibration had no effect on the output of the fiber optic pressure sensor.

10.3.3 Dynamic Testing

One of the potential advantages of fiber optic sensors over conventional pressure sensors is faster dynamic response. The transmission of the pressure signal to the remote electronics utilizes light beams which makes it almost

Internal Temperature Lag

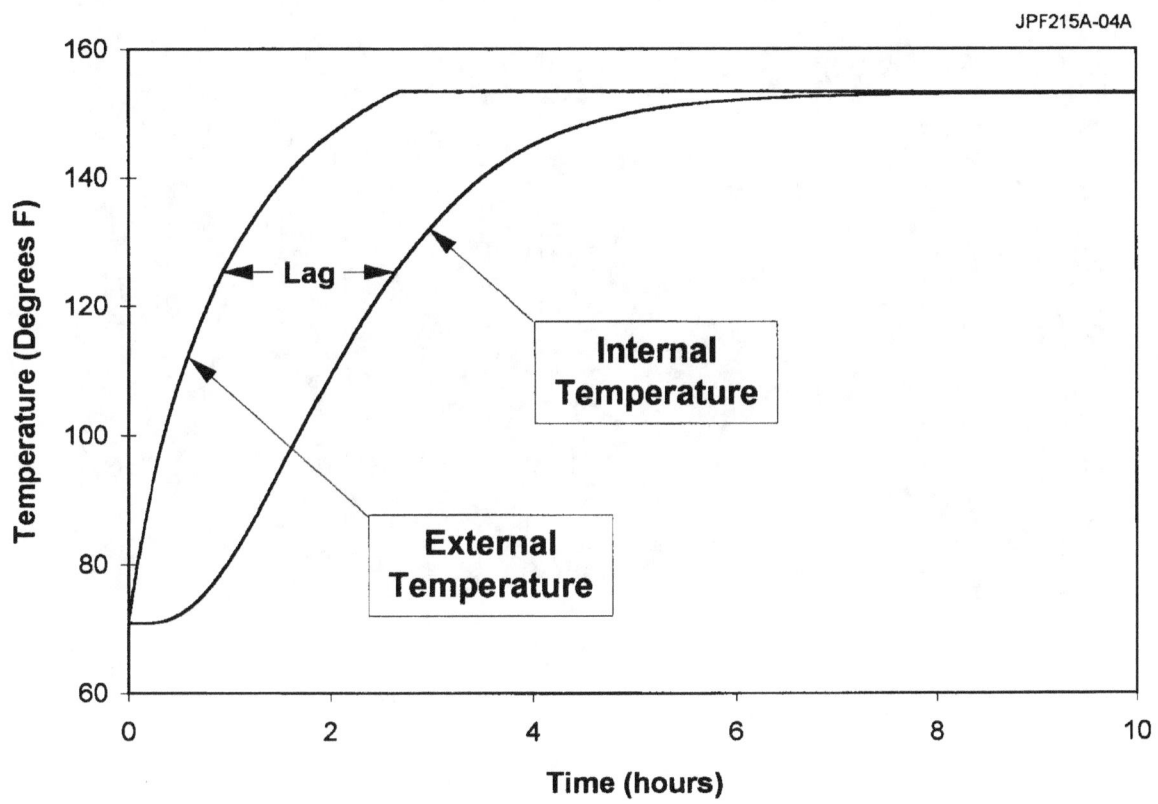

Figure 10.8   Temperature Lag Between the External and Internal
Temperature Seen by the Fiber Optic Pressure Transducer

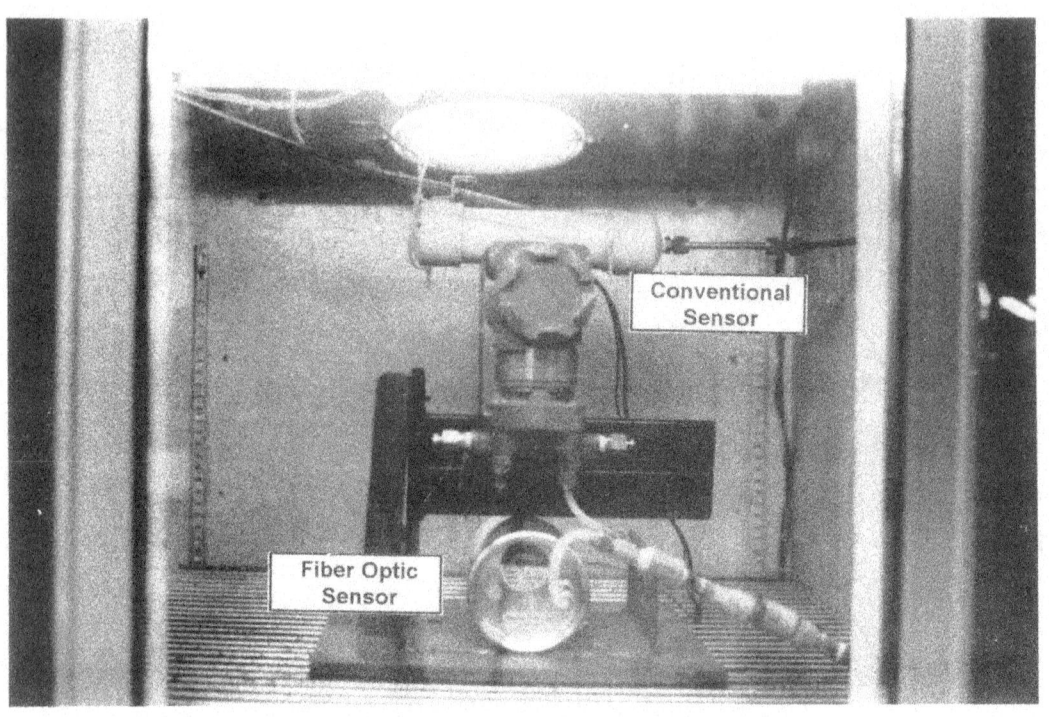

**Figure 10.9   Photograph of the Test Setup for a Series of Temperature
Cycling Tests Performed in the Environmental Chamber**

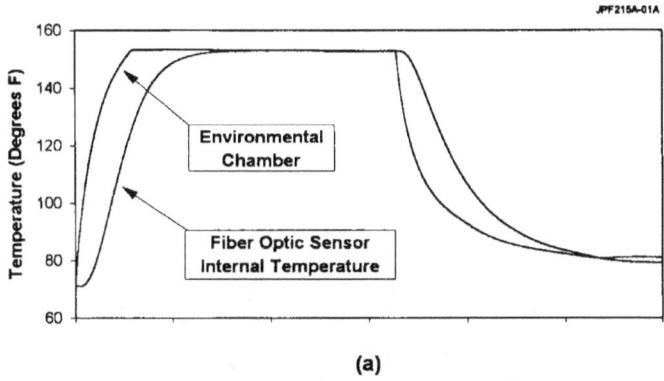

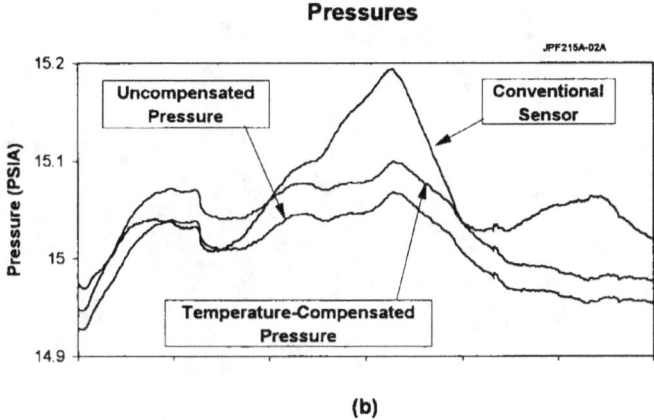

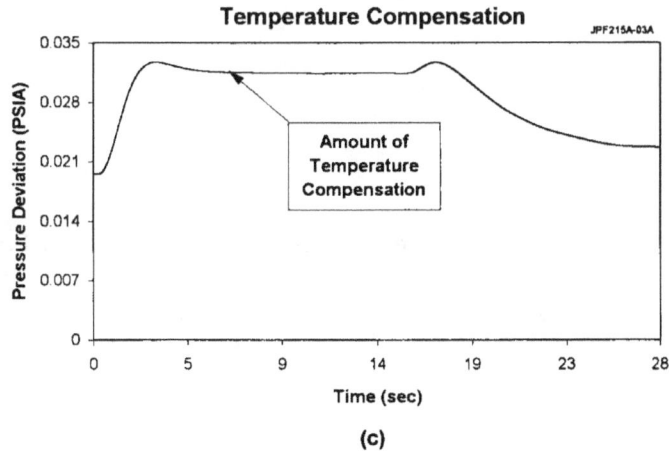

Figure 10.10   Results of One Series of Temperature Cycling Tests

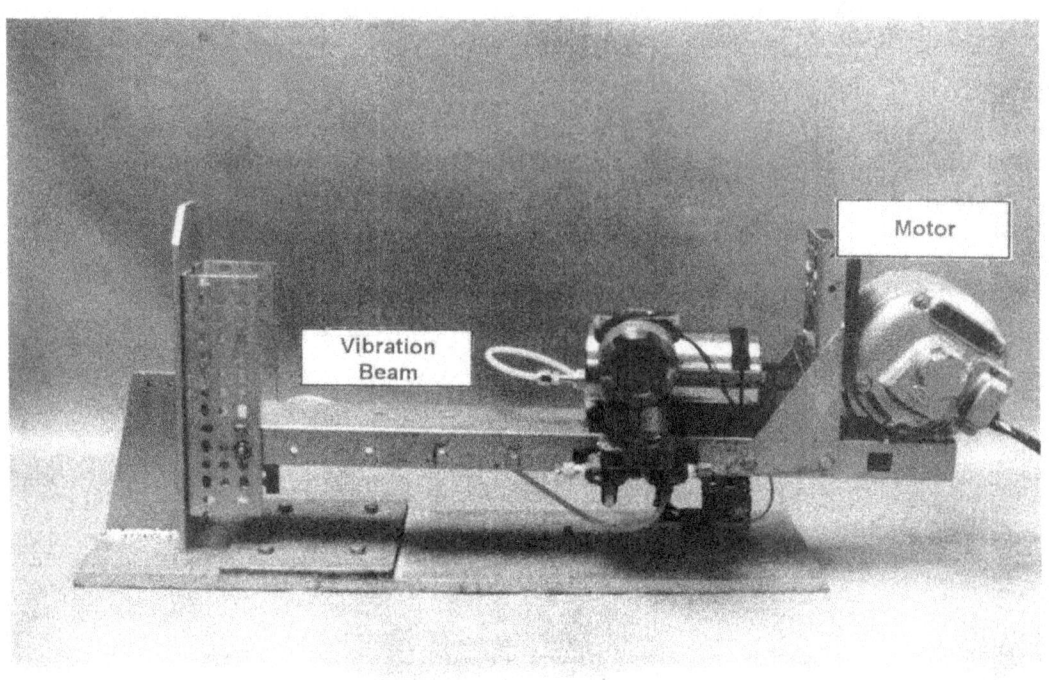

Figure 10.11   Photograph of the Test Setup for Vibration Tests

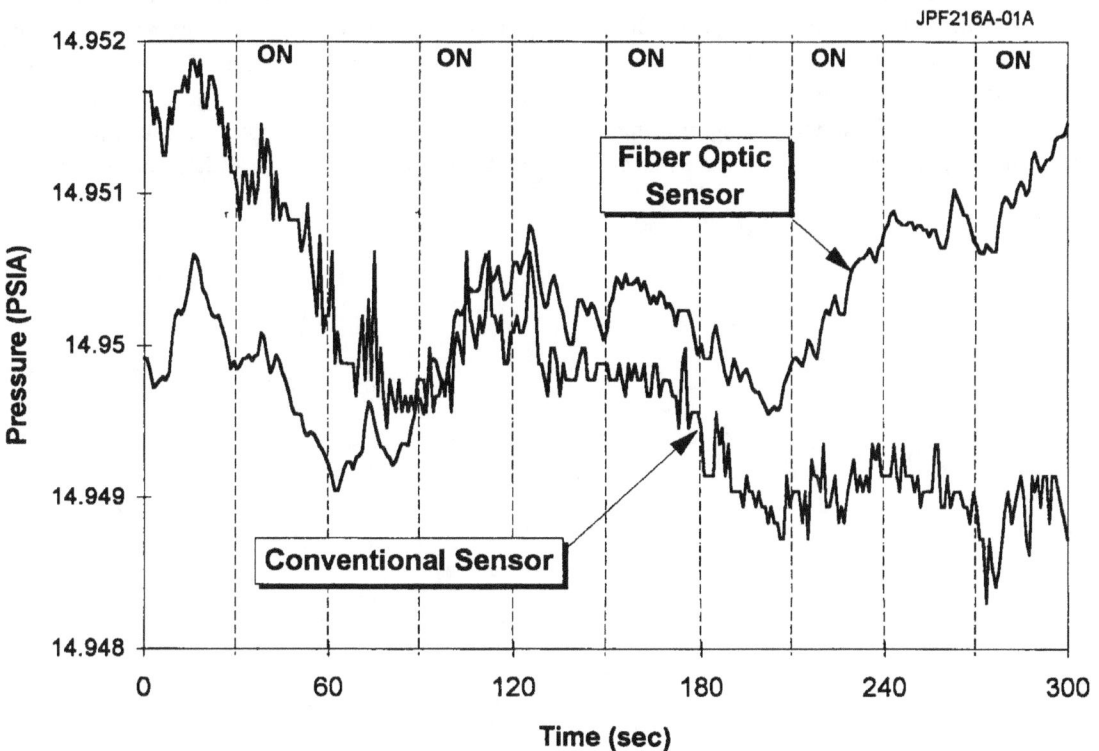

Figure 10.12   Results of One Series of Vibration Tests Showing no Visible Effect on the Fiber Optic Sensor Output as the Vibration Beam Motor was Cycled On and Off

instantaneous. Therefore, the limiting factors are the sensing elements that convert the process measurement to a mechanical displacement and the remote electronics that convert the incoming light signal to a measured pressure. The dynamic testing performed on the Paroscientific sensor attempted to establish its dynamic response characteristics.

Figure 10.13 illustrates the test setup for the ramp testing performed on the fiber optic pressure sensor. As seen in this figure, the pressure ramp was applied to both the fiber optic sensor and a pressure transmitter produced by the Validyne Engineering Corporation which was used as a reference sensor. The Validyne transmitter is a high-speed reference transmitter with a response time of less than 10 milliseconds. A computer initiated the ramp test by triggering a solenoid and then acquired data from both the reference sensor and the fiber optic sensor. A manually-controlled needle valve was used to control the rate of the pressure ramp.

Figure 10.14(a) shows the results from one ramp test while 10.14(b) shows the same results zoomed in on the first part of the transient. As seen in this figure, the dynamic response of the fiber optic pressure sensor is faster than that of the Validyne transmitter.

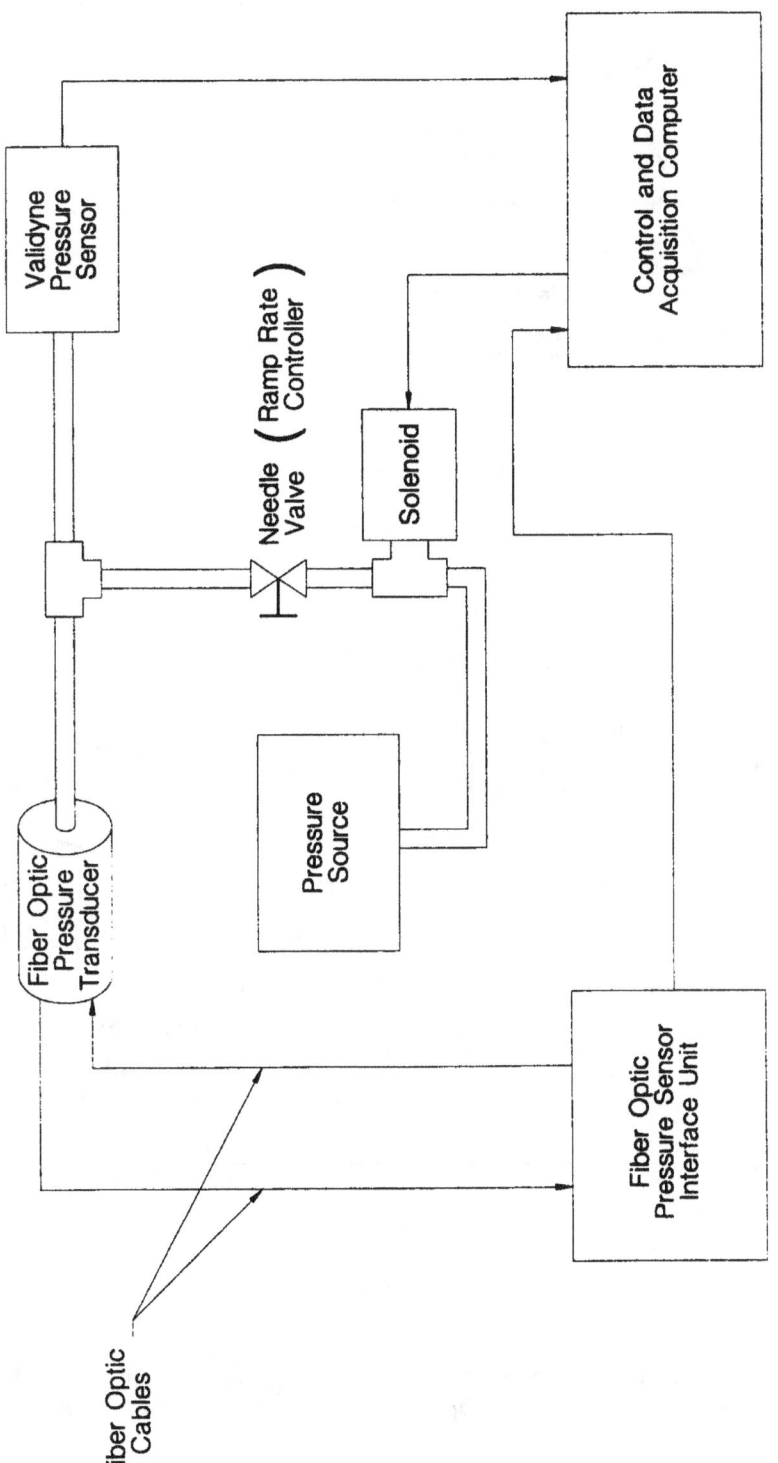

Figure 10.13   Illustration of the Test Setup for Dynamic Ramp Testing

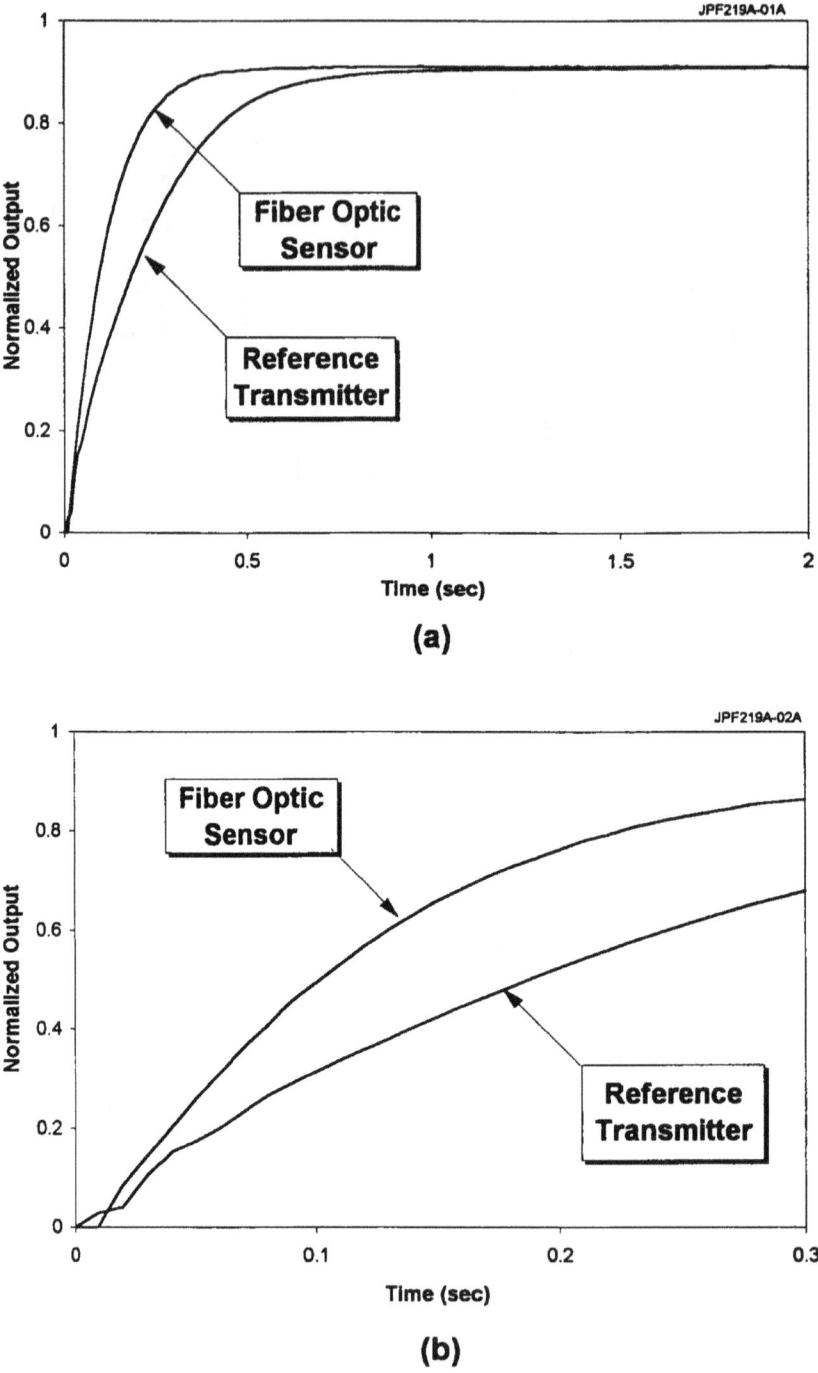

Figure 10.14   Results of an Increasing Pressure Ramp Showing the Faster Dynamic Response of the Fiber Optic Pressure Sensor versus the Reference Sensor

11. CONCLUSIONS

The results of a six-month Phase I research project funded by the Office of Nuclear Regulatory Research of the U.S. Nuclear Regulatory Commission are documented in this report. The purpose of this effort was to establish the state-of-the-art in fiber optic sensing and determine if these sensors can be used for safety-related applications in nuclear power plants. The study included experimental work involving a fiber optic pressure sensor that was tested at the AMS laboratory. Furthermore, an informal survey of over one hundred fiber optic sensor manufacturers, researchers, authors, and others was performed using questionnaires that were sent out and followed by telephone contacts. These efforts in addition to an extensive literature review have led to the conclusion that although fiber optic pressure sensors have several advantages over the conventional pressure sensors, they are not presently ready for use in nuclear power plants because:

1) Fiber optic sensing technologies are still evolving and their performance has not been extensively demonstrated in harsh environments.

2) Fiber optic pressure sensors are not readily available for industrial applications; there is only a handful of manufacturers which produce these sensors mostly as special order items for specific applications in medical, aerospace, chemical, and automotive industries.

3) Fiber optic pressure sensors have not been qualified according to the IEEE standards that are used for qualification testing of Class 1E equipment for nuclear power plants.

4) There are questions about the ability of fiber optic pressure sensors to serve in nuclear radiation environments.

5) Fiber optic pressure sensors are much more expensive than conventional sensors.

REFERENCES

1. *Fiber Optics I*, Instrument Society of America (ISA), Research Triangle Park, North Carolina, 1988.

2. Saleh, B.E.A., and Teich, M.C., *Fundamentals of Photonics*, John Wiley & Sons, Inc., New York, New York, 1991.

3. Udd, E., *Fiber Optic Sensors: an Introduction for Engineers and Scientists*, John Wiley & Sons, Inc., New York, New York, 1991.

4. Glomb, W.L., "Electro-Optic Architecture (EOA) for Sensors and Actuators in Aircraft Propulsion Systems," NASA Report CR-182270, June 1989.

5. Leonberger, F.J., Glomb, W.L., Dunphy, J.R., "Fly by Light: Fiber Optics For Aircraft Communication, Control and Sensing," *Optical Fiber Communication Conference*, 1990 Technical Digest Series, Vol. 1, Optical Society of America, p.48, January 22-26, 1990.

6. Murphy, K.A., et al., "Single Mode Variable-Sensitivity Fiber Optic Sensors," NASA Report CR-190492, 1992.

7. Reed, S.E., et al., "Fiber Optic Total Pressure Transducer for Aircraft Applications," Presented at the *Society of Photo-Optical Instrumentation Engineers Symposium on Optical Tools for Manufacturing and Advanced Automation*, Boston, Massachusetts, September 7-10, 1993.

8. National Science Foundation, Research Initiative Announcement and Description -- "Sensors and Sensor Systems for Power Systems and other Dispersed Civil Infrastructure Systems," Arlington, Virginia, January 1993.

9. Electric Power Research Institute, "Fiber-Optic Sensors," RP8004 Technical Brief, January 1993.

10. Shepard, R.L., and Thacker, L.H., "Evaluation of Pressure Sensing Concepts: A Technology Assessment," ORNL/TM-12296, Oak Ridge National Laboratory, September 1993.

11. Weiss, J.M., "Power Plant Fiber-Optic Sensors Being Developed," EPRI Journal, p. 5, April/May 1994.

12. Valenti, M., "A New Generation of Power Plant Sensors," *Mechanical Engineering*, pp. 54-58, September 1994.

13. Berthold, J.W., "Overview of Fiber-Optic Intensity Sensors for Industry," Presented at the *Society of Photo-Optical Instrumentation Engineers O-E/FIBERS 1987 Conference*, August 16-21, 1987.

14. Miller, D.W., et al., "An Experimental Performance Assessment of Currently Available Optical Fibers In Nuclear Reactor Radiation Environments," Presented at the *Nuclear Plant Instrumentation, Control and Man-Machine Interface Technologies*, Oak Ridge, TN, April 1993.

15. Jacobson, C.P., "Fiber Optic Sensors for Navy Ships," *SPIE*, Vol. 2072, pp. 6-11, February 1994.

16. Hegner, H.R., and Whitesel, H.K., "Study of Fiber Optic Sensor Reliability, Durability and Failure Modes for Shipboard Machinery," *SPIE*, Vol. 2072, pp. 12-21, February 1994.

17. Hashemian, H.M., et al., "Long Term Performance and Aging Characteristics of Nuclear Plant Pressure Transmitters," NUREG/CR-5851, U.S. Nuclear Regulatory Commission, Washington, DC, March 1993.

18. Hashemian, H.M., et al., "Validation of Smart Sensor Technologies for Instrument Calibration Reduction in Nuclear Power Plants," NUREG/CR-5903, U.S. Nuclear Regulatory Commission, Washington, DC, January 1993.

19. Krohn, D.A., *Fiber Optic Sensors - Fundamentals and Applications, Second Edition*, Instrument Society of America (ISA), Research Triangle Park, North Carolina, 1992.

20. Lin, H., and Ho, C., "Optical Pressure Transducer," *Rev. Sci. Instrum.*, Vol. 64, No. 7, pp. 1999-2002, July 1993.

21. Kersey, A.D., and Dandridge, A., "Distributed and Multiplexed Fiber-Optic Sensors," *Optical Fiber Sensors*, 1988 Technical Digest Series, Vol. 2, Optical Society of America, Washington, D.C., pp. 60-71, 1988.

22. Murphy, K.A., et al., "Quadrature Phase-Shifted, Extrinsic Fabry-Perot Optical Fiber Sensors," *Optic Letters*, Vol. 16, No. 4, pp. 273-275, February 15, 1991.

23. Claus, R.O., et al., "Extrinsic Fabry-Perot Sensor For Strain and Crack Opening Displacement Measurements From - 200 to 900 °C," *Smart Mater. Struct.*, Vol. 1, pp. 237-242, 1992.

24. Beheim, G., Fritsch, K., and Poorman, R.N., "Fiber-Linked Interferometric Pressure Sensor," *Rev. Sci. Instrum.*, Vol. 58, No. 9, pp. 1655-1659, September 1987.

25. Gustafsson, K., and Hök, B., "A Fibre-Optic Pressure Sensor in Silicon Based on Fluorescence Decay," *Sensors and Actuators*, 19, pp. 327-332, 1989.

26. Georgopoulos, C.J., *Fiber Optics and Optical Isolators*, Don White Consultants, Inc., Gainesville, Virginia, 1982.

27. Culshaw, B., "Fiber-Optic Sensors - A Brief Review," *Sensors & Actuators, Optical Transducers -- Proceedings of the S&A Symposium of the University of Twente Enschede, The Netherlands*, pp. 13-30, November 15-16, 1990.

28. Ferdinand, P., et al., "The Potential for Distributed Sensors and Optical Fibre Sensor Networks In The Electric Power Industry," *Meas. Sci. Technol.*, pp. 908-916, 1990.

29. Thornton, K.E.B., Uttamchandani, D., and Culshaw, B., "Optically Excited Micromechanical Resonator Pressure Sensor," *Optical Fiber Sensors*, 1988 Technical Digest Series, Vol. 2, Optical Society of America, Washington, D.C., pp. 433-436, 1988.

30. Weiss, J.M., Shepard, R.L., "Evaluation of Advanced Pressure Sensor Technology," Paper #92-0642, Instrument Society of America, Research Triangle Park, North Carolina, 1992.

31. Spenner, K., "Microbending Pressure and Displacement Sensor," *Third International Conference on Optical Fiber Sensors*, OSA/IEEE, San Diego, CA, pp. 146-148, February 1985.

32. Rao, Y.J., and Jackson, D.A., "Prototype Fiber-Optic-Based Ultrahigh Pressure Remote Sensor With Built-In Temperature Compensation," *Rev. Sci. Instrum.*, Vol. 65, No. 5, p. 1695-1698, May 1994.

33. Bouzid, A., Abushagur, A.G., and He, Z., "O-Ring Fiber Optic Pressure Sensor," *Optical Engineering*, Vol. 33, No. 4, pp. 1074-1077, April 1994.

34. Ohkawa, M., Izutsu, M., and Sueta, T., "Integrated Optic Pressure Sensor on Silicon Substrate," *Applied Optics*, Vol. 28, No. 23, pp. 5153-5157, December 1, 1989.

35. Haiming, X., "Birefringent Fiberoptic Pressure Sensor with High Sensitivity," *Chinese Physics*, Vol. 9, No. 3, pp. 842-847, July-Sept., 1989.

36. Berthold, J.W., Ghering, W.L., and Varshneya, D., "Design and Characterization of A High-Temperature, Fiber-Optic Pressure Transducer," *Journal of Lightwave Technology*, Vol. LT-5, No. 7, July 1987.

37. McCollum, T., and Spector, G.B., "Fiber Optic Microbend Sensor for Detection of Dynamic Fluid Pressure at Gear Interfaces," *Rev. Sci. Instrument.*, Vol. 65, No. 3, pp. 724-729, March 1994.

38. Berthold, J.W., "Status and Review of Fiber-Optic Sensors in Industry," *SPIE*, Vol. CR-44, 1992.

39. Barwicz, A., and Bock, W.J., "An Electronic High-Pressure Measuring System Using a Polarimetric Fiber-Optic Sensor," *IEEE Transactions on Instrumentation and Measurement*, Vol. 39, No. 6, pp. 976-981, December 1990.

40. Iwamoto, K., and Kamata, I., "Pressure Sensor Using Optical Fibers," *Applied Optics*, Vol. 29, No. 3, pp. 375-378, January 20, 1990.

41. Berthold, J.W., "Field Test Results on Fiber-Optic Pressure Transmitter System," *SPIE*, Vol. 1584, 1991.

42.   Nakayama, T., and Kyuma, K., "Fiber-Optic Sensors in Japan," *Third International Conference on Optical Fiber Sensors*, Optical Society of America, San Diego, California, p. 118, February 13-14, 1985.

43.   *Fiber Optics II*, Instrument Society of America (ISA), Research Triangle Park, North Carolina, 1989.

44.   Hocker, G.B., "Fiber-Optic Sensing of Pressure and Temperature," *Applied Optics*, Vol. 18, No. 9, pp. 1445-1448, May 1, 1979.

45.   Dakin, J.P., Wade, C.A., Withers, P.B., "An Optical Fibre Pressure Sensor," *SPIE Vol. 74 Fibre Optics 87: Fifth International Conference on Fibre Optics and Opto-Electronics*, pp. 194-201, 1987.

46.   Rao, Y.J., Culshaw, B., and Uttamchandani, D., "An Improved TDM System Applied to the Multiplexing of Silicon Microresonator Sensors Exhibiting Identical Characteristics," Proceedings of the *8th Optical Fiber Sensors Conference*, Monterey, California, pp. 270-273, January 29-31, 1992.

47.   Korsah, K., Clark, R.L., and Holcomb, D.E., "A Methodology for Evaluating "New" Technologies in Nuclear Power Plants," Paper # 94-2689, Instrument Society of America (ISA), Research Triangle Park, North Carolina, 1994.

48.   Holcomb, D.E, Miller, D.W., and Weiss, J.M., "An Experimental Performance Assessment of Currently Available Optical Fibers in Nuclear Reactor Radiation Environments," Paper #92-0643, Instrument Society of America (ISA), Research Triangle Park, North Carolina, 1992.

49.   Institute of Electrical and Electronics Engineers, IEEE Standard 323-1974, "IEEE Standard for Qualifying Class 1E Equipment for Nuclear Power Generating Stations."

50.   Institute of Electrical and Electronics Engineers, IEEE Standard 344-1975, "IEEE Recommended Practices for Seismic Qualification of Class 1E Equipment for Nuclear Power Generating Stations."

51.   Paros, J.M., "Fiber Optic Resonator Pressure Tranducers," *Measurements and Control,* Issue 154, pp. 144-148, September 1992.

52.   Paroscientific, Inc., "Digiquartz® Precision Pressure Instruments -- Programming and Operation Manual," Redmond, Washington, 1993.

APPENDIX A

BIBLIOGRAPHY

BIBLIOGRAPHY

Fiber Optic Sensing and Related Texts

Berthold, J.W., et al., Editors, *Proc. SPIE Volume 2072 - Fiber Optic Physical Sensors in Manufacturing and Transportation*, SPIE - The International Society for Optical Engineering, Bellingham, Washington, 1994.

Dakin, J., and Culshaw, B., *Optical Fiber Sensors: Principles and Components*, Artech House, Inc., Boston, Massachusetts, 1988.

Georgopoulos, C.J., *Fiber Optics and Optical Isolators*, Don White Consultants, Inc., Gainesville, Virginia, 1982.

ISA, *Fiber Optics I*, Research Triangle Park, North Carolina, 1988.

ISA, *Fiber Optics II*, Research Triangle Park, North Carolina, 1989.

Krohn, D.A., *Fiber Optic Sensors - Fundamentals and Applications, Second Edition*, ISA, 1992.

Laurin Publishing Co., Inc., *The Photonics Directory*, Volumes 1-4, Pittsfield, Massachusetts, 1994.

Palais, J., *Fiber Optic Communications*, Prentice-Hall, Inc., Englewood Cliffs, New Jersey, 1984.

Paul, D.K., *Fiber Optics Reliability and Testing*, SPIE Optical Engineering Press, Boston, Massachusetts, 8-9 September 1993.

Saleh, B.E.A., and Teich, M.C., *Fundamentals of Photonics*, John Wiley & Sons, Inc., New York, New York, 1991.

Udd, E., *Fiber Optic Sensors: an Introduction for Engineers and Scientists*, John Wiley & Sons, Inc., New York, New York, 1991.

Zhilin, V.G., *Optical-Fiber Velocity and Pressure Transducers*, Hemisphere Publishing Corporation, New York, New York, 1990.

Fiber Optic Pressure Sensor Papers

Barwicz, A., and Bock, W.J., "An Electronic High-Pressure Measuring System Using a Polarimetric Fiber-Optic Sensor," *IEEE Transactions on Instrumentation and Measurement*, Vol. 39, No. 6, pp. 976-981, December 1990.

Beheim, G., Fritsch, K., and Poorman, R.N., "Fiber-Linked Interferometric Pressure Sensor," *Rev. Sci. Instrum.*, Vol. 58, No. 9, pp. 1655-1659, September 1987.

Berthold, J.W., "Field Test Results on Fiber-Optic Pressure Transmitter System," *SPIE*, Vol. 1584, 1991.

Berthold, J.W., Ghering, W.L., and Varshneya, D., "Design and Characterization of A High-Temperature, Fiber-Optic Pressure Transducer," *Journal of Lightwave Technology*, Vol. LT-5, No. 7, July 1987.

Bock, W.J., Beaulieu, M., and Domanski, A.W., "GaAs-Based Fiber-Optic Pressure Sensor," *IEEE Transactions on Instrumentation and Measurement*, Vol. 41, No. 1, pp. 68-71, February, 1992.

Bock, W.J., and Eftimov, T.A., "Simultaneous Hydrostatic Pressure and Temperature Measurement Employing an LP_{01}-LP_{11} Fiber-Optic Polarization-Sensitive Intermodal Interferometer," *IEEE Transactions on Instrumentation and Measurement*, Vol. 43, No. 2, pp. 337-340, April 1994.

Bock, W.J., Wisniewski, R., and Wolinski, T.R., "Fiber-Optic Strain-Gauge Manometer Up to 100 MPA," *IEEE Transactions on Instrumentation and Measurement*, Vol. 41, No. 1, pp. 71-76, February, 1992.

Bock, W.J., Wolinski, T.R., and Barwicz, A., "Development of a Polarimetric Optical Fiber Sensor for Electronic Measurement of High Pressure," *IEEE Transactions on Intrumentation and Measurement*, Vol. 19, No. 5, pp. 715-721, 1990.

Bouzid, A., Abushagur, A.G., and He, Z., "O-Ring Fiber Optic Pressure Sensor," *Optical Engineering*, Vol. 33, No. 4, pp. 1074-1077, April 1994.

Cho, Y.C., and Soderman, P.T., "Fiber-Optic Interferometric Sensors for Measurements of Pressure Fluctuations: Experimental Evaluation," NASA Technical Memorandum 104002, January 1993.

Dabkiewicz, P., and Jansen, K., "Fiber-Optic Pressure Sensor Using Birefringence in Side-Hole Fiber," *SPIE Vol. 734 Fibre Optics 87: Fifth International Conf. on Fibre Optics and Opto-Electronics*, pp. 202-206, 1987.

Dakin, J.P., Wade, C.A., Withers, P.B., "An Optical Fibre Pressure Sensor," *SPIE Vol. 74 Fibre Optics 87: Fifth International Conference on Fibre Optics and Opto-Electronics*, pp. 194-201, 1987.

Garcia-Valenzuela, A., and Tabib-Azar, M., "Fiber-Optic Force and Displacement Sensor Based on Speckle Detection With 0.1nN and 0.1 Å Resolution," *Sensors and Actuators A*, pp. 199-208, 1993.

Grossman, B., et al., "Fiberoptic Pore Water Pressure for Civil Engineering Applications," *SPIE*, Vol. 2072, 1994.

Gustafsson, K., and Hök, B., "A Fibre-Optic Pressure Sensor in Silicon Based on Fluorescence Decay," *Sensors and Actuators*, 19, pp. 327-332, 1989.

Haiming, X., "Birefringent Fiberoptic Pressure Sensor with High Sensitivity," *Chinese Physics*, Vol. 9, No. 3, pp. 842-847, July-Sept., 1989.

He, G., and Wlodarczyk, M.T., "Laboratory and In-Vehicle Evaluations of Fiber Optic Combustion Pressure Sensor," *SPIE*, Vol. 2072, 1994.

He, G., and Wlodarczyk, M.T., "Spark Plug-Integrated Fiber Optic Combustion Pressure Sensor," Proceedings of the *Sensors Expo*, pp. 211-216, September 29 - October 1, 1992.

Henderson, P.J., Spencer, J., and Jones, G.R., "Pressure Sensing Using A Chromatically Addressed Diaphragm," *Meas. Sci. Technol.*, pp. 88-94, 1993.

Hocker, G.B., "Fiber-Optic Sensing of Pressure and Temperature," *Applied Optics*, Vol. 18, No. 9, pp. 1445-1448, May 1, 1979.

Ikeda, M.H., Sun, M.H., and Phillips, S.R., "Piezoelectric Crystal Based Fiberoptic Pressure Sensor," *Optical Fiber Sensors*, 1988 Technical Digest Series, Vol. 2, Optical Society of America, Washington, D.C., pp. 172-178, 1988.

Iwamoto, K., and Kamata, I., "Pressure Sensor Using Optical Fibers," *Applied Optics*, Vol. 29, No. 3, pp. 375-378, January 20, 1990.

Kawabata, Y., Yasunaga, K., and Ishibashi, N., "Photoacoustic Spectrometry Using a Fiber-Optic Pressure Sensor," *Analytical Chemistry*, Vol. 65, No. 23, pp. 3493-3496, December 1, 1993.

Leonard, M., "Tiny Optical Sensor Takes Temperature, Pressure Measurements," *Electronic Design*, Vol. 37, No. 18, pp. 20 & 22, August 24, 1989.

Libo, Y., and Anping, Q. "Fiber-Optic Diaphragm Pressure Sensor With Automatic Intensity Compensation," *Sensors and Actuators A.*, Vol. 28, pp. 29-33, 1991.

Libo, Y., Shunling, R., and Jian, P., "Automatic Compensation Fiber-Optic Differential Pressure Sensor," *Sensors and Actuators A.*, Vol. 36, pp. 183-185, 1993.

Lin, H., and Ho, C., "Optical Pressure Transducer," *Rev. Sci. Instrum.*, Vol. 64, No. 7, pp. 1999-2002, July 1993.

Martens, G., Kordts, J., and Weidlinger, G., "Loss-Compensated Photoelastic Fiber-Optic Pressure Sensor," *Applied Optics*, Vol. 28, No. 23, pp. 5149-5152, December 1, 1989.

McCollum, T., and Spector, G.B., "Fiber Optic Microbend Sensor for Detection of Dynamic Fluid Pressure at Gear Interfaces," *Rev. Sci. Instrument.*, Vol. 65, No. 3, pp. 724-729, March 1994.

Mortensen, P., "Rubber Fiber Senses Pressure," *Laser Focus World*, Vol. 25, No. 5, p. 146, May 1989.

Murphy, M.M., and Jones, G.R., "Optical Fibre Pressure Measurement," *Meas. Sci. Technol.*, pp. 258-262, 1993.

Ohkawa, M., Izutsu, M., and Sueta, T., "Integrated Optic Pressure Sensor on Silicon Substrate," *Applied Optics*, Vol. 28, No. 23, pp. 5153-5157, December 1, 1989.

Paroscientific, Inc., "Quartz Transducers for Precision Under Pressure," *Mechanical Engineering*, Vol. 109, No. 5, May 1987.

Paros, J.M., "Fiber-Optic Pressure Sensors with 0.01% Accuracy," Publication Status Unknown.

Rao, Y.J., and Jackson, D.A., "Prototype Fiber-Optic-Based Fizeau Medical Pressure Sensor That Uses Coherence Reading," *Optics Letters*, Vol. 18, No. 24, pp. 2163-2155, December 15, 1993.

Rao, Y.J., and Jackson, D.A., "Prototype Fiber-Optic Based Pressure Probe With Built-In Temperature Compensation With Signal Recovery by Coherence Reading," *Applied Optics*, Vol. 32, No. 34, pp. 7110-7113, December 1, 1993.

Rao, Y.J., and Jackson, D.A., "Prototype Fiber-Optic-Based Ultrahigh Pressure Remote Sensor With Built-In Temperature Compensation," *Rev. Sci. Instrum.*, Vol. 65, No. 5, p. 1695-1698, May 1994.

Reed, S.E., et al., "Fiber Optic Total Pressure Transducer for Aircraft Applications," Presented at the *Society of Photo-Optical Instrumentation Engineers Symposium on Optical Tools for Manufacturing and Advanced Automation*, Boston, Massachusetts, September 7-10, 1993.

Sato, R., "Pressure Balancing Structure for Fiber-Optic Flexural Disk Acoustic Sensor," *Jpn. J. Appl. Phys.*, Vol. 32, pp. 2473-2476, 1993.

Shepard, R.L., and Thacker, L.H., "Evaluation of Pressure Sensing Concepts: A Technology Assessment," ORNL/TM-12296, Oak Ridge National Laboratory, September 1993.

Spenner, K., "Microbending Pressure and Displacement Sensor," *Third International Conference on Optical Fiber Sensors*, OSA/IEEE, San Diego, CA, pp. 146-148, February 1985.

Thornton, K.E.B., Uttamchandani, D., and Culshaw, B., "Optically Excited Micromechanical Resonator Pressure Sensor," *Optical Fiber Sensors*, 1988 Technical Digest Series, Vol. 2, Optical Society of America, Washington, D.C., pp. 433-436, 1988.

Tran, T.A, et al., "Surface-Mounted Optical Fiber Sensors for Measurement of Hypersonic Boundary Layer Instability Modes," Proceedings of the *Second International Conference on Intelligent Materials*, pp. 1242-1247, 1994.

Varshneya, D., Ghering, W.L., and Berthold, J.W., "High-Temperature Fiber-Optic Microbend Pressure Sensor," *Third International Conference on Optical Fiber Sensors*, OSA/IEEE, San Diego, CA, pp. 140, February 1985.

Voet, M.R.H., Desforges, F.X., and Barel, A.R.L., "An Optical Fiber Network for Analog Temperature and Pressure Sensing Purposes," Proceedings of the *8th Optical Fiber Sensors Conference*, Monterey, California, pp. 205-208, January 29-31, 1992.

Wearn, R., "Fiber-Optic Quartz Crystal Pressure Transducers," *Sensors*, September, 1992.

Weiss, J.M., "Fiber Optics-Based Pressure Sensors," *EPRI ERL*, p. 7, February 1994.

Weiss, J.M., Shepard, R.L., "Evaluation of Advanced Pressure Sensor Technology," ISA Paper # 92-0642, 1992.

Wesson, L.N., et al., "Fiber-Optic Pressure Sensor System for Gas Turbine Engine Control," Publication Status Unknown.

Wlodarczyk, M.T., et al., "Fiber Optic Pressure Sensor for Combustion Monitoring and Control," *SPIE*, Vol. 1587, 1991.

Wlodarczyk, M.T., and He, G., "A Fiber-Optic Combustion Pressure Sensor System for Automotive Engine Control," *Sensors*, pp. 35-42, June 1994.

Wolthuis, R., et al., "Development of a Dual Function Sensor System For Measuring Pressure and Temperature at the Tip of a Single Optical Fiber," *IEEE Transactions on Biomedical Engineering*, Vol. 40, No. 3, pp. 298-302, March 1993.

Wu, S., Yin S., and Yu., F.T.S., "Sensing With Fiber Specklegrams," *Applied Optics*, Vol. 30, No. 31, pp. 4468-4470, November 1, 1991.

Fiber Optic Sensors and Applications Papers

Berthold, J.W., "Overview of Fiber-Optic Intensity Sensors for Industry," Presented at the *Society of Photo-Optical Instrumentation Engineers O-E/FIBERS 1987 Conference*, August 16-21, 1987.

Berthold, J.W., "Status and Review of Fiber-Optic Sensors in Industry," *SPIE*, Vol. CR-44, 1992.

Bock, W.J., "Automatic Calibration of a Fiber-Optic Strain Sensor Using a Self-Learning System," *IEEE Transactions on Instrumentation and Measurement*, Vol. 43, No. 2, pp. 341-346, April 1994.

Boutacoff, D., "The Story in Brief," *EPRI Journal*, pp. 15-21, September 1990.

Bunch, R.M. "Optical Fiber Sensor Experiments for the Undergraduate Physics Laboratory," *Am. J. Phys.*, Vol. 58, No. 9, pp. 870-874, September 1990.

Claus, R.O., "Optical Fiber Sensors For Nondestructive Evaluation," *ASNT Spring Conference*, Nashville, TN, March 31, 1993.

Claus, R.O., et al., "Extrinsic Fabry-Perot Sensor For Strain and Crack Opening Displacement Measurements From - 200 to 900 °C," *Smart Mater. Struct.*, Vol. 1, pp. 237-242, 1992.

Covington, C.E., Blake, J., and Carrara, S.L.A., "Two-Mode Fiber-Optic Bending Sensor with Temperature and Strain Compensation," *Optic Letters*, Vol. 19, No. 9, pp. 676-678, May 1, 1994.

Culshaw, B., "Fiber-Optic Sensors - A Brief Review," *Sensors & Actuators, Optical Transducers* -- Proceedings of the S&A Symposium of the University of Twente Enschede, The Netherlands, pp. 13-30, November 15-16, 1990.

Day, G.W., et al., "Faraday Effect Sensors for Magnetic Field and Electric Current," *SPIE Interfermetry '94*, Warsaw, Poland, May 16-20, 1994.

Diemeer, M.B.J., "Fiber-Optic Microbend Sensors: Sensitivity As A Function of Distortion Wavelength and Deformation Force," *Sensors and Actuators* -- Proceedings S&A Symposium of the Twente University of Technology, Enschede, The Netherlands, pp. 29-36, November 1-2, 1984.

Dubowski, J.J., Lebecki, K., and Buchanan, M., "Fiber Optic CdMnTe Magnetic Field Sensor Made by the Laser Ablation Deposition Technique," *IEEE Transactions on instrumentation and Measurements*, Vol. 42, No. 2, pp. 322-325, April 1994.

Edwards, J.G., "Integrated Optic Sensors for Industry," Proceedings of the *Sensors Expo*, Chicago, Illinois, pp. 1-11, September 29 - October 1, 1992.

Electric Power Research Institute, "Fiber-Optic Sensors," RP8004 Technical Brief, January 1993.

Elliott, T.C., "Special Report: Advanced Sensors," *Power*, August 1994.

Falco, L., Parriaux, O., "Structured Metal Coatings for Distributed Fiber Sensors," Proceedings of the *8th Optical Fiber Sensors Conference*, Monterey, California, pp. 254-257, January 29-31, 1992.

Ferdinand, P., et al., "The Potential for Distributed Sensors and Optical Fibre Sensor Networks In The Electric Power Industry," *Meas. Sci. Technol.*, pp. 908-916, 1990.

Fiber & Electro-Optics Research Center, "Datasheet," Virginia Tech, Blacksburg, VA.

Fiber & Electro-Optics Research Center, *LIGHTNEWS*, Virginia Tech, Blacksburg, VA, Fall/Winter, 1993.

Glomb, W.L., "Electro-Optic Architecture (EOA) for Sensors and Actuators in Aircraft Propulsion Systems," NASA Report CR-182270, June 1989.

Greene, J.A., et al., "Elliptical-Core Two-Mode Fiber Sensors and Devices Incorporating Photoinduced Refractive Index Gratings," NASA Report CR-190495, 1992.

Harmer, A.L., "Fibre Optic Sensors For Sale ?" *Optical Fiber Sensors*, 1988 Technical Digest Series, Vol. 2, Optical Society of America, Washington, D.C., pp. 2&3, 1988.

He, G., and Cuomo, F.W., "Displacement Response, Detection Limit, and Dynamic Range of Fiber-Optic Lever Sensors," *Journal of Lightwave Technology*, Vol. 9, No. 11, pp. 1618-1625, November 1991.

Hegner, H.R., and Whitesel, H.K., "Study of Fiber Optic Sensor Reliability, Durability and Failure Modes for Shipboard Machinery," *SPIE*, Vol. 2072, pp. 12-21, February 1994.

Herzog, J.P., Roth, P., and Meyrueis, P., "Optical Fiber Flowmeter with Temperature Correction," *Sensors and Actuators A.*, pp. 219-223, 1991.

Holcomb, D.E., and Antonescu, C., "A Review of Potential Uses for Fiber Optic Sensors in Nuclear Power Plants, With Attendant Benefits in Plant Safety and Operational Efficiency," Presented at the *21st Water Reactor Safety Information Meeting*, Bethesda, Maryland, October 25-27, 1993.

Holcomb, D.E, Miller, D.W., and Weiss, J.M., "An Experimental Performance Assessment of Currently Available Optical Fibers in Nuclear Reactor Radiation Environments," ISA Paper # 92-0643, 1992.

Ikeda, M.H., Sun, M.H., and Phillips, S.R., "Fiberoptic Flow Sensor," *Optical Fiber Sensors*, 1988 Technical Digest Series, Volume 2, Part 2, Optical Society of America, Wahington, D.C., pp. 438-445, January 27-29, 1988.

Iwamoto, K., and Kamata, I., "Liquid-Level Sensor With Optical Fibers," *Applied Optics*, Vol. 31, No. 1, pp. 51-54, January 1, 1992.

Jackson, D.A., "Minitutorial: Fiber-Optic Sensors," *Third International Conference on Optical Fiber Sensors*, Optical Society of America, San Diego, California, p. 118, February 13-14, 1985.

Jacobson, C.P., "Fiber Optic Sensors for Navy Ships," *SPIE*, Vol. 2072, pp. 6-11, February 1994.

Johnson, C., and Brittain, T., "Fiber Optics Reduce Data Errors," *Chemical Engineering*, August 1992.

Kersey, A.D., "Fiber Sensors and Measurements," Technical Digest of the *Optical Fiber Communication Conference*, San Diego, California, p. 81, February 18-22, 1991.

Kersey, A.D., and Berkoff, T.A., "Fiber-Optic Bragg Grating Strain Sensor with Drift-Compensated High-Resolution Interferometric Wavelength-Shift Detection," *Optic Letters*, Vol. 18, No. 1, pp. 72-74, January 1, 1993.

Kersey, A.D., and Dandridge, A., "Distributed and Multiplexed Fiber-Optic Sensors," *Optical Fiber Sensors*, 1988 Technical Digest Series, Vol. 2, Optical Society of America, Washington, D.C., pp. 60-71, 1988.

Korsah, K., Clark, R.L., and Holcomb, D.E., "A Methodology for Evaluating "New" Technologies in Nuclear Power Plants," ISA Paper # 94-2689, 1994.

Kovacs, M.P., "Fiberoptic Sensors Approach Commercial Success," *Laser Focus World*, pp. 175-183, March 1991.

Leonberger, F.J., Glomb, W.L., Dunphy, J.R., "Fly by Light: Fiber Optics For Aircraft Communication, Control and Sensing," *Optical Fiber Communication Conference*, 1990 Technical Digest Series, Vol. 1, Optical Society of America, p.48, January 22-26, 1990.

Linderner, D.K., and Claus, R.O., "Optical Fiber Sensors for Materials and Structures Characterization," NASA Report CR-194090, September 1991.

McNeil M., and Landis, C., "Fiber-Optic Switches in Carrageenin Manufacture," *Sensors*, September 1993.

Miller, D.W., et al., "An Experimental Performance Assessment of Currently Available Optical Fibers In Nuclear Reactor Radiation Environments," Presented at the *Nuclear Plant Instrumentation, Control and Man-Machine Interface Technologies*, Oak Ridge, TN, April 1993.

Murphy, K.A., et al., "Extrinsic Fabry-Perot Optical Fiber Sensor," Proceedings of the *8th Optical Fiber Sensors Conference*, Monterey, California, pp. 193-196, January 29-31, 1992.

Murphy, K.A., et al., "Fabry-Perot Fiber-Optic Sensors in Full-Scale Fatigue Testing on an F-15 Aircraft," *Applied Optics*, Vol. 31, No. 4, pp. 431-433, February 1, 1992.

Murphy, K.A., et al., "Quadrature Phase-Shifted, Extrinsic Fabry-Perot Optical Fiber Sensors," *Optic Letters*, Vol. 16, No. 4, pp. 273-275, February 15, 1991.

Murphy, K.A., et al., "Single Mode Variable-Sensitivity Fiber Optic Sensors," NASA Report CR-190492, 1992.

Nakayama, T., and Kyuma, K., "Fiber-Optic Sensors in Japan," *Third International Conference on Optical Fiber Sensors*, Optical Society of America, San Diego, California, p. 118, February 13-14, 1985.

Naqwi, A.A., and Petrik, S., "Fiber-Optic Dual-Cylindrical Wave Sensor for Measurement of Wall Velocity Gradient In A Fluid Flow," *Applied Optics*, Vol. 32, No. 30, pp. 6128-6131, October 20, 1993.

National Science Foundation, Research Initiative Announcement and Description -- "Sensors and Sensor Systems for Power Systems and other Dispersed Civil Infrastructure Systems," Arlington, Virginia.

Orrell, P., "Fibre Optic Sensors," *Engineering*, pp. 15 & 16, December 1993.

Pandya, D., "Fiberoptic Sensors Find Growing Niche In World Market," *Laser Focus World*, pp. 65 & 66, August 1989.

Poole, S.B., "Optical-Fiber Sensors Attract Attention in Australia," *Fiberoptics News and Markets*, pp. 160, 161, 164.

Poole, S.B., "Application Specific Optical Fibres and Fibre Devices for Optical Fibre Sensors," Proceedings of the *8th Optical Fiber Sensors Conference*, Monterey, California, pp. 274-278, January 29-31, 1992.

Rao, Y.J., Culshaw, B., and Uttamchandani, D., "An Improved TDM System Applied to the Multiplexing of Silicon Microesonator Sensors Exhibiting Identical Characteristics," Proceedings of the *8th Optical Fiber Sensors Conference*, Monterey, California, pp. 270-273, January 29-31, 1992.

Saaski, E., "Fiberoptic Sensors, Measure in Adverse Conditions Using a New Fiberoptic Technology," *Measurements & Control*, February 1989.

Saaski, E., and Har, "Fiber-Optic Fabry-Perot Temperature Sensors," *Temperature*, Vol. 6, Part 2, pp. 731-734, 1992.

Sigel, G.H., "Minitutorial: Fiber-Optic Sensors," *Third International Conference on Optical Fiber Sensors*, Optical Society of America, San Diego, California, p. 84, February 11-13, 1985.

Sigel, G.H., "Optical Fiber Sensors," Technical Digest of the *Optical Fiber Communication Conference and Sixth International Conference on Integrated Optics and Optical Fiber Communication*, Reno, Nevada, p. 70, January 19-22, 1987.

Taylor, E.W., "Principles of Photonics," Presented at the *Instrumentation Society of America Conference*, Baltimore, MD, May 5, 1994.

Tran, T.A., et al., "Stabilized Extrinsic Fiber-Optic Fizeau Sensor for Surface Acoustic Wave Detection," *Journal of Lightwave Technology*, Vol. 10, No. 10, pp. 1499-1506, October 1992.

Valenti, M., "A New Generation of Power Plant Sensors," *Mechanical Engineering*, pp. 54-58, September 1994.

Wang, A., and Murphy, K.A., "Optical-Fiber Temperature Sensor Based on Differential Spectral Reflectivity," *Smart Mater. Struct.*, Vol. 1, 1991.

Weir, K., et al., "A Fibre Optic, Low Coherence Laser Doppler Anemometer System for Determining Flow Velocity," Proceedings of the *8th Optical Fiber Sensors Conference*, Monterey, California, pp. 237-240, January 29-31, 1992.

Weiss, J.M., "Power Plant Fiber-Optic Sensors Being Developed," EPRI Journal, p. 5, April/May 1994.

Weiss, J.M., Esselman, W., and Lee, R., "Assess Fiberoptics Sensors for Key Powerplant Measurements," *Power*, pp. 55-58, October 1990.

Woracek, D., "Fiber Optic Sensors for Microwave Oven and Other Electrically Noisy Process Control Environments," Publication Status Unknown.

Wu, Y., et al., "Fiber-Optic Ultrasonic Sensor Using Raman-Nath Light Diffraction," *IEEE Transactions on Ultrasonics, Ferroelectrics and Frequency Control*, Vol. 41, No. 2, pp. 166-171, March 1994.

Yoshino, Y., "Heterodyne Technology for Optical Sensors," *Optical Fiber Sensors*, 1988 Technical Digest Series, Vol. 2, Optical Society of America, Washington, D.C., pp. 40-43, 1988.

Fiber Optic Product Bulletins & Manufacturers' Literature

3M, "Fiber Optic Cable 200/230 for Industrial Communications," 3M Data Sheet, 1993.

3M, "Series 6000 Modular Fiber Optic Modem," 3M Product Bulletin, 1993.

Applied Measurement Technologies, Babcock & Wilcox, "AMTEC Publications," Alliance, OH.

Applied Measurement Technologies, Babcock & Wilcox, "Current List of AMTEC PRODUCTS AND PROTOTYPES," Alliance, OH.

Applied Measurement Technologies, Babcock & Wilcox, "Development of a Long-Term, Post-Closure Radiation Monitoring System," Alliance, OH.

Applied Measurement Technologies, Babcock & Wilcox, "Fiber Optic Void Fraction Sensor," Alliance, OH.

Applied Measurement Technologies, Babcock & Wilcox, "Pressure Transducer That Performs In Hostile Environments."

McCarter-Carr, "Fiber Optics," Catalog #99.

Photonetics, "Fiberoptic Absorbance Measurement Probes," Wakefield MA.

Photonetics, "Fiberoptic Analytical Instruments," Wakefield, MA.

Photonetics, "Fiberoptic Pressure Sensor Metallic Probes," Wakefield, MA.

Photonetics, "Fiberoptic Pressure Sensor Non-Conductive Probes," Wakefield, MA.

Photonetics, "Fiberoptic Temperature Sensor Metallic Probes," Wakefield, MA.

Photonetics, "Fiberoptic Temperature Sensor Non-Conductive Probes," Wakefield, MA.

Photonetics, "Fiberoptic Transmission/Reflection Measurement Cells," Wakefield MA.

Photonetics, "Fiberoptic Refractive Index Sensor Probes," Wakefield, MA.

Photonetics, "Fiberoptic Sensor Systems for Two-Phase Flows," Wakefield, MA.

Weed Fiber-Optics, "CCTR Series 8 Channel Contract Closure Data Link," Round Rock, TX.

Weed Fiber-Optics, "EOS-FM Series Intrinsically Safe Control Module," Round Rock, TX, October 1992.

Weed Fiber-Optics, "FOTR Series Analog Fiber Optic Data Links," Round Rock, TX.

Weed Fiber-Optics, "Fiber Optic Liquid Level Switch," Round Rock, TX.

Weed Fiber-Optics, "Light Talk," Volume 1, Number 1, Round Rock, TX.

Weed Fiber-Optics, "Light Talk," Volume 1, Number 2, Round Rock, TX.

Weed Fiber-Optics, "Light Talk," Volume 1, Number 3, Round Rock, TX.

Weed Fiber-Optics, "Light Talk," Volume 1, Number 4, Round Rock, TX.

Weed Fiber-Optics, "MUX64 Series 64 Channel Analog/Digital Fiber Optic Multiplexer," Round Rock, TX, July 19, 1991.

Weed Fiber Optics, "Pushbutton Components," Round Rock, TX.

Weed Fiber-Optics, "Rugged RS 232 Fiber Optic Modems Asynchronous," Round Rock, TX.

Weed Fiber-Optics, "Series 8000 4 Channel Multiplexer," Round Rock, TX.

APPENDIX B

SURVEY AND INTERVIEW QUESTIONNAIRES

INFORMAL SURVEY QUESTIONNAIRE

SURVEY OF FIBER OPTIC PRESSURE SENSOR MANUFACTURERS

Name / Title: _____

Organization: _____

Date questionnaire completed: _____ Conducted By_____
(If telephone survey)

Section 1
(For all Fiber Optic Sensor Manufacturers)

1.   Do you manufacture fiber optic pressure sensors?   Yes _____ No _____

      If yes, what type?

      ___ Diaphragm deflection (Extrinsic)
      ___ Diaphragm deflection (Intrinsic)
      ___ Fabry-Perot interferometry
      ___ Mach-Zehnder interferometry
      ___ Piezoluminescent
      ___ Other _____

2.   What other types of fiber optic sensors do you manufacture?   (Check all
      that apply)

      ___ Temperature
      ___ Strain
      ___ Displacement
      ___ Flow
      ___ Sound
      ___ Bending
      ___ Void Fraction
      ___ Other

3.   Do you have a specification sheet for your sensor(s)?   Yes _____ No _____

      If yes, please include a copy with your survey response.

      (skip 4-11 as appropriate, if you are including a specification sheet)

Fiber Optic Sensor Survey

4. What is the operating temperature range for your sensors?

5. What is the operating relative humidity range for your sensor(s)?

6. What is the measurement range of your sensor(s)?

7. What is the resolution of the digital (or analog) output?

8. What is the calibrated accuracy of your sensor(s)?

9. Is the calibration accuracy temperature dependent?   Yes _____ No _____

10. Does your sensor(s) have a drift specification?

11. What is the recommended calibrating interval?

12. How many sensors are supported by each transmitter (electronics)?

13. What is the range of prices for your sensor elements? For your transmitters?

Section 2
(For Fiber Optic Sensor Manufacturers that Supply
Fiber Optic Pressure Sensors)

14. Do you manufacture sensors capable of measuring water pressure?

15. Do you manufacture sensors capable of measuring fluid level?

16. Do you manufacture sensors capable of measuring fluid flow?

17. Have you determined the response time of your sensors?   Yes _____ No _____

    If yes, what is the range? _____ Sec.

Fiber Optic Sensor Survey

18. Have you determined the response time of your
    sensor/transmitter system?　　　　　　　　　Yes _____ No _____

19. If no, is that something you would like to know?　　Yes _____ No _____

20. Do you perform any other tests which we have not asked about?

21. For what additional characteristics do you test?

22. Have you identified any failure modes in your sensors/transmitters or in similar
    sensors/transmitters from other manufacturers?

23. Is your company interested in providing pressure transmitters to the nuclear
    industry?

24. Would you be interested in having a sample of your transmitters tested at no
    cost?

25. Do you have any non-proprietary test data, papers or
    evaluations that you could release for publications?　　Yes _____ No _____

INTERVIEW QUESTIONNAIRE

FIBER OPTIC PRESSURE SENSOR TELEPHONE INTERVIEW

Name / Title: _____

Organization: _____

Date questionnaire completed: _____ Conducted By _____
(If telephone survey)

1. What is your involvement, if any, with fiber optics?

2. Do you make, design, test, perform research on, or use fiber optic sensors? Also, if so, what types of sensors.

3. Do you make, design, test, perform research on, or use fiber optic pressure sensors? Also, if so, what types of pressure sensors.

4. What is the state of the art in fiber optic pressure sensing?

5. What is the most popular use (or measurand) for fiber optic sensors?

6. What are the advantages and disadvantages of fiber optic sensors over conventional sensors? What are the specific advantages and disadvantages of fiber optic pressure sensors over conventional pressure sensors?

7. What are the markets for fiber optic sensors, in general (where are they used)? What are the markets for fiber optic pressure sensors?

8. Do you perceive a current lack of availability of fiber optic pressure sensors? If so, then what are the reasons for this lack of availability?

9. Do fiber optic pressure sensors have the potential to be used for safety-related applications in nuclear power plants?

10. What type of fiber optic pressure sensors has the best potential for nuclear plant use? Please explain.

Fiber Optic Sensor Interview

11. What type of research and development work is currently in progress in this area?

12. What type of research and development can AMS perform that would be beneficial to industry?

13. Can you provide a fiber optic pressure sensor for testing in this research project, or can you refer us to anyone else who might be able to provide a sensor?

14. Can you provide lab testing data, or pictures or diagrams, etc. or refer us to another source of information?

NRC FORM 335
(2-89)
NRCM 1102,
3201, 3202

U.S. NUCLEAR REGULATORY COMMISSION

BIBLIOGRAPHIC DATA SHEET

(See instructions on the reverse)

1. REPORT NUMBER (Assigned by NRC, Add Vol., Supp., Rev., and Addendum Numbers, If any.)
NUREG/CR-6312

2. TITLE AND SUBTITLE

Assessment of Fiber Optic Pressure Sensors

3.	DATE REPORT PUBLISHED	
	MONTH	YEAR
	April	1995

4. FIN OR GRANT NUMBER
W6315

5. AUTHOR(S)

H.M. Hashemian, C.L. Black, J.P. Farmer

6. TYPE OF REPORT

Technical

7. PERIOD COVERED *(Inclusive Dates)*

7/1/94 – 1/31/95

8. PERFORMING ORGANIZATION – NAME AND ADDRESS *(If NRC, provide Division, Office or Region, U.S. Nuclear Regulatory Commission, and mailing address; if contractor, provide name and mailing address.)*

Analysis and Measurement Services Corporation
AMS 9111 Cross Park Drive
Knoxville, TN 37923

9. SPONSORING ORGANIZATION – NAME AND ADDRESS *(If NRC, type "Same as above"; if contractor, provide NRC Division, Office or Region, U.S. Nuclear Regulatory Commission, and mailing address.)*

Division of Systems Technology
Office of Nuclear Regulatory Research
U.S. Nuclear Regulatory Commission
Washington, D.C. 20555-0001

10. SUPPLEMENTARY NOTES

11. ABSTRACT *(200 words or less)*

The principle of operation of fiber optic pressure sensors and the potential of
these sensors for use in nuclear power plants are described in this report.
Also included is a review of current research on fiber optic sensing
technologies, a comparison of fiber optic pressure sensors with conventional
pressure sensors, a discussion on advantages and disadvantages of fiber optic
pressure sensors, a review of failure modes of these sensors, and results of a
survey of fiber optic sensor manufacturers.

12. KEY WORDS/DESCRIPTORS *(List words or phrases that will assist researchers in locating the report.)*

Fiber Optic Sensors
Nuclear Power Plants
Pressure Sensors
Instrumentation
Advanced Reactors

13. AVAILABILITY STATEMENT

Unlimited

14. SECURITY CLASSIFICATION

(This Page)

Unclassified

(This Report)

Unclassified

15. NUMBER OF PAGES

16. PRICE

NRC FORM 335 (2-89)

Printed on recycled paper

Federal Recycling Program